March 2, 1982

Yuji Kodama

LIE-BÄCKLUND TRANSFORMATIONS IN APPLICATIONS

SIAM Studies in Applied Mathematics

This series of monographs focuses on mathematics and its applications to problems of current concern to industry, government, and society. These monographs will be of interest to applied mathematicians, numerical analysts, statisticians, engineers, and scientists who have an active need to learn useful methodology for problem solving.

Robert L. Anderson
Nail H. Ibragimov

LIE-BÄCKLUND TRANSFORMATIONS IN APPLICATIONS

siam *Philadelphia / 1979*

Library of Congress Catalog Card Number: 78–78207

Contents

Preface

SIAM Studies in Applied Mathematics focuses on the presentation of mathematical theory and its applications in the context of formulating and solving concrete physical problems. Areas of mathematics are selected which are highly motivated by physical applications in the belief that they possess a high potential for future fruitful development. Further, these areas clearly exhibit the vital interdependence of the development of mathematics with that of science and technology. It is our belief that we have identified one such area in this first volume in the *Studies*.

We gratefully acknowledge Professor L. V. Ovsjannikov's scientific influence on our investigations in this area and his support of this work. Our thanks also go to Professors W. F. Ames, M. Flato, and R. Rączka for their valuable discussions and encouragement.

This material was prepared during reciprocal exchange visits to each other's countries. In this regard, we especially wish to thank Professor L. V. Ovsjannikov, Director of the Institute of Hydrodynamics, USSR Academy of Sciences, Siberian Branch, and Professor C. J. Hand, Academic Vice President of the University of the Pacific, as well as acknowledge the financial support of the American and Soviet Academies of Sciences and the Fulbright–Hays Committee through the Council for International Exchange of Scholars. In addition, this work was partially supported by NSF Grant OIP 74-01416 (Special Foreign Currency Program).

We would like to thank Professor W. F. Ames, who invited us to undertake the project; Mrs. Marilyn Koch, Mrs. Sandy Rux, and Mrs. Carol Sarnoff for their excellent typescript and invaluable help during the preparation of this manuscript; and SIAM for their assistance and cooperation.

We dedicate this monograph to Lois and Galia as a token of our thanks for their support and understanding of this venture.

ROBERT L. ANDERSON
NAIL H. IBRAGIMOV

Introduction

During the past ten years there has been a resurgence in interest in the analysis of differential equations and their solutions from the point of view of their invariance properties under a type of surface transformation known as a Bäcklund transformation. One of the purposes of this monograph is to provide an introduction to the classical treatment, as established primarily in the papers of S. Lie and A. V. Bäcklund, of these and general surface transformations as higher-order tangent transformations. In this context we discuss classical as well as recent examples of Bäcklund transformations as applied to nonlinear optics (sine–Gordon), nonlinear waves (Kortweg–de Vries and Liouville), a turbulence model (Burgers) and quantum mechanics (nonlinear Schrödinger). Since techniques of constructing these transformations are presented in detail, the volume will be of considerable use to the scientist and engineer concerned with analysis of mathematical models of physical phenomena.

A second purpose is to present recent results which establish the group theoretical context of a generalization of Lie's first-order tangent (contact) transformation groups and its application to differential equations. We call this generalization a Lie–Bäcklund tangent transformation group. These Lie–Bäcklund transformation groups have application to equations which describe the time evolution of systems encountered in engineering, hydrodynamics, mechanics, physics, control mechanisms, ecology, economics, and biochemistry. Here we discuss examples of these transformations from the areas of mechanics, gas dynamics, hydrodynamics, relativity, and quantum mechanics.

While the monograph presupposes some prior knowledge of graduate analysis and group theory, the exposition is self-contained and readily penetrable. The detailed examples amplify the material and demonstrate how other models can be analyzed. Sufficient references, both theoretical and applied, provide a variety of supplementary information—including references to original and fundamental articles of the nineteenth century.

The selection of material in Chapter 1 is predicated on our wish to present in one place the fundamental ideas, notions, and results of the classical papers of

Lie, Bäcklund, and Bianchi, as well as exhibit their own acknowledged mutual influence in developing the foundations of this subject. Another of our purposes will be served if readers will be persuaded to read for themselves this classical literature.

The elaboration of the structure of these surface transformations is contained in the classical literature with the exception of recent consideration of groups of Lie–Bäcklund transformations and their projections on integral surface manifolds of differential equations. These groups and their associated structure are treated in Chapter 2.

Chapter 3 contains some applications chosen from our work to illustrate the general structure discussed in Chapters 1 and 2. Professor W. F. Ames's invaluable contributions to this monograph (Chapter 1, §§7, 8; and Chapter 4) were written to provide examples of applications of Bäcklund transformations. The bibliographical references for these sections appear as footnotes to the text; they are distinct from the References section to be found on pages 119–121.

Chapter 1

Classical Foundations

In this chapter we present the classical treatment of surface transformations, which is founded on the idea of higher-order tangent transformations. The original papers of S. Lie [1], [3] and A. V. Bäcklund [1], [2], [3], [4], which develop this treatment, evidence the mutual influence each had on the other's contribution. Their work on surface transformations was the result of a search for a generalization of Lie's theory of first-order tangent (contact) transformations and its application to differential equations. This problem was also set in the context of one of the central problems occurring in the classical literature—namely the investigation of the reduction of the problem of the integration of an arbitrary differential equation to the corresponding problem for a linear differential equation(s).

The basic ideas underlying the possibility and importance of applying higher-order tangent transformations to differential equations were clearly formulated by Lie [1] as two questions in his 1874 paper. The program of realizing this generalization was undertaken by A. V. Bäcklund [1], [2] in his 1874 and 1876 papers, where he considered finite-order tangent transformations. His main result in these papers is that no single-valued surface transformations exist other than Lie tangent transformations.

The multifaceted possibilities offered by these general surface transformations for the study of differential equations were revealed in Lie's 1880 treatment of surfaces of constant curvature (Lie [3], [4]). In these papers, he constructed the first nontrivial example of such transformations treating Bianchi's geometrical construction (Bianchi [1]) as an ∞-valued transformation of surfaces of constant curvature. Lie's analytical expression of this transformation is given by four particular equations relating two sets of surface elements (x, y, z, p, q) and (x', y', z', p', q'). Further, he showed that his ∞-valued transformation has the property that it is a surface transformation on surfaces of constant curvature only. Another feature of this transformation, which distinguishes it from the transformations considered by Bäcklund in his 1874 and 1876 papers, is that it conserves the order of the second-order differential equation defining surfaces of constant curvature; in fact, it leaves this equation invariant. These specific properties led him to pose the general problem of determining the ∞-valued transformations which are surface transformations only on integral surfaces of a given differential equation and which leave this differential equation invariant.

Bäcklund generalized this example by extending Lie's expression for Bianchi's construction to transformations which are given by four arbitrary functions between sets of surface elements. In particular, he stated the consistency conditions for the overdetermined system of differential equations which result when one treats these four equations as a transformation of a given surface into a surface. This type of transformation was called in the classical literature a *Bäcklund transformation*, and that name is still applied to precisely this type of surface transformation. Goursat [1] and Clairin [1] investigated the characterization of those transformations which are surface transformations on integral surfaces of second-order partial differential equations. Goursat [2] later clarified the statement of this problem for transformations of families of surface elements.

Although here we call general surface transformations, *Lie–Bäcklund transformations*, we shall adopt the standard nomenclature when referring to special types of these transformations, namely, Lie point and Lie tangent transformations as well as the Bäcklund transformations mentioned in the preceding paragraph.

I. SURFACE TRANSFORMATIONS

§1. Lie's First Question

The general concept of a surface transformation that we will consider in this monograph has its origin in the notion of first-order tangent transformations, also called contact transformations. In fact, it appeared in the classical literature as a direct extension of the notion of first-order tangent transformations to higher-order tangency. Because of the fundamental role played, both historically and logically, by first-order tangent transformations in the conceptualization of general surface transformations, we review here aspects of Lie's theory of these transformations (Lie [2], [5]). We follow convention and refer to first-order tangent transformations as *Lie tangent transformations*.

Consider the space $\mathbb{R}^{2n+1}$ of variables $x=(x^1,\cdots,x^n)$, u, $\underset{1}{u}=(u_1,\cdots,u_n)$ and an invertible transformation T:

$$(1.1)\qquad \begin{aligned} x'&=f\left(x, u, \underset{1}{u}\right),\\ u'&=\phi\left(x, u, \underset{1}{u}\right),\\ \underset{1}{u}'&=\underset{1}{\psi}\left(x, u, \underset{1}{u}\right),\end{aligned}$$

in this space, where $f=(f^1,\cdots,f^n)$ and $\underset{1}{\psi}=(\psi_1,\cdots,\psi_n)$. The action of the transformation T is extended to new variables—"differentials" $dx=(dx_1,\cdots,dx_n)$, du, $d\underset{1}{u}=(du_1,\cdots,du_n)$—according to formulas

$$
\begin{aligned}
dx'^i &= \frac{\partial f^i}{\partial x^j}dx^j + \frac{\partial f^i}{\partial u}du + \frac{\partial f^i}{\partial u_j}du_j, \\
du' &= \frac{\partial \phi}{\partial x^j}dx^j + \frac{\partial \phi}{\partial u}du + \frac{\partial \phi}{\partial u_j}du_j, \qquad (1.2)\\
du_i' &= \frac{\partial \psi_i}{\partial x^j}dx^j + \frac{\partial \psi_i}{\partial u}du + \frac{\partial \psi_i}{\partial u_j}du_j,
\end{aligned}
$$

so that the combined action of (1.1) and (1.2) is a transformation $\tilde{T}$ in the prolonged $(x, u, \underset{1}{u}, dx, du, d\underset{1}{u})$-space. The transformation $\tilde{T}$ is called the *prolongation* of T.

DEFINITION 1.1. A transformation T is called a *Lie tangent transformation* if the first-order tangency condition

$$du - u_j\,dx^j = 0 \qquad (1.3)$$

is invariant with respect to the prolonged transformation $\tilde{T}$.

In this analytical definition, transformations T of the form (1.1) are considered as point transformations in the $(x, u, \underset{1}{u})$-space. This analytical treatment of Lie tangent transformations also has a clear geometrical sense. In order to illustrate this, consider the case where $n=1$. According to Definition 1.1 any geometrical configuration in the (x, u)-space consisting of a curve C and its tangent τ at a point $P \in C$ (Fig. 1) is converted by a Lie tangent transformation

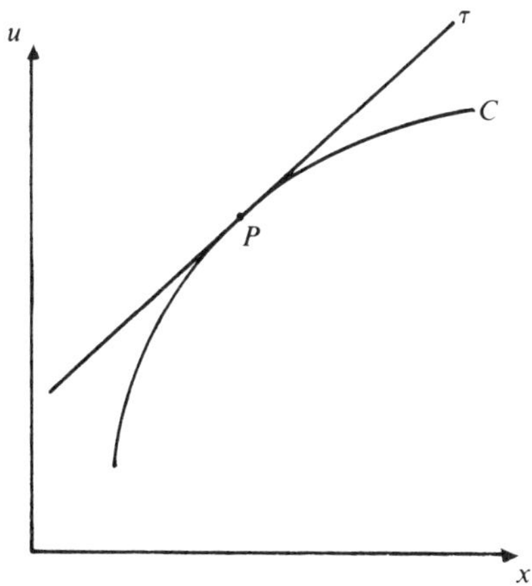

FIG. 1. A curve C and its tangent τ at the point P.

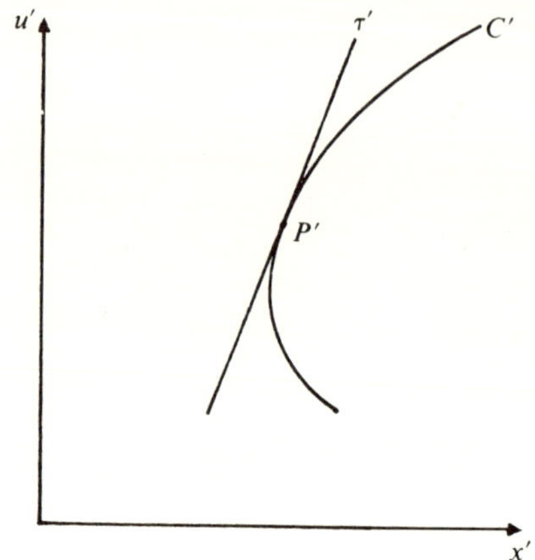

FIG. 2. The image of the geometrical configuration shown in Fig. 1 under a Lie tangent transformation.

T into a similar geometrical configuration (represented by Fig. 2) in (x', u')-space.

This fact leads directly to the possibility of transforming differential equations by means of (1.1). More specifically, given a differential equation, say a first-order partial differential equation

$$\omega\left(x, u, \underset{1}{u}\right)=0, \tag{1.4}$$

one can by means of the action of (1.1) transform (1.4) into a similar equation (without raising the order of the original equation)

$$\omega'\left(x', u', \underset{1}{u'}\right)=0, \tag{1.5}$$

where the left-hand side is defined by the equality

$$\omega'\left(f\left(x, u, \underset{1}{u}\right), \phi\left(x, u, \underset{1}{u}\right), \psi\left(x, u, \underset{1}{u}\right)\right)=\omega\left(x, u, \underset{1}{u}\right). \tag{1.6}$$

Now, given a solution (integral surface) of (1.4), then (1.1), according to the geometrical meaning of Lie tangent transformations, converts these integral-surfaces into integral-surfaces of (1.5). This means that a Lie tangent transformation (1.1) maps a first-order partial differential equation in (x, u)-space into another or the same first-order partial differential equation in (x', u')-space. The action of the transformation (1.1) can be extended to include second and/or any higher-order derivatives through the operations of differentiation and elimination, e.g.,

$$\underset{2}{u'}=\underset{2}{\psi}\left(x, u, \underset{1}{u}, \underset{2}{u}\right),$$

where

$$\underset{2}{u}=\{u_{ij}|i,j=1,\cdots,n\} \quad \text{and} \quad \underset{2}{\psi}=\{\psi_{ij}|i,j=1,\cdots,n\}.$$

Again, the transformation (1.1) together with these natural prolongations converts a second- and/or higher-order partial differential equation into a similar one (without raising the order of the original equation). Lie realized that a generalization of the concept of the tangent transformation could be important in applications to second and higher-order differential equations. He formulated this idea in his 1874 paper as the first of two questions (Lie [1], p. 223).

Lie's First Question. Are there transformations which are not first-order tangent transformations and for which tangency of higher order is an invariant condition?

Lie predicted a negative answer to this question. From Lie's group-theoretical treatment of transformations, it is natural here to understand transformations as invertible maps. With this understanding of transformations as invertible ones (or, in classical terminology, as single-valued surface transformations), one can interpret the main results in Bäcklund's first papers (Bäcklund [1], [2]) as a verification of Lie's conjecture. This question is discussed in §§2, 3.

§2. Finite-Order Generalization

Consider the space of variables $(x, u, \underset{1}{u},\cdots, \underset{k}{u})$, where

$$x=(x^1,\cdots,x^n)\in\mathbb{R}^n, \qquad u\in\mathbb{R},$$

$$\underset{s}{u}=\{u_{i_1\cdots i_s}|i_1,\cdots,i_s=1,\cdots n\}, \qquad 1\leqq s\leqq k,$$

and an invertible transformation T:

$$\begin{aligned} x'&=f\left(x, u, \underset{1}{u},\cdots, \underset{k}{u}\right),\\ u'&=\phi\left(x, u, \underset{1}{u},\cdots, \underset{k}{u}\right),\\ \underset{1}{u}'&=\underset{1}{\psi}\left(x, u, \underset{1}{u},\cdots, \underset{k}{u}\right),\\ &\cdots\cdots\cdots\cdots\cdots\\ \underset{k}{u}'&=\underset{k}{\psi}\left(x, u, \underset{1}{u},\cdots, \underset{k}{u}\right). \end{aligned} \tag{2.1}$$

Here we have employed notation similar to that used in §1, namely, $f=(f^1,\cdots,f^n)$ and $\underset{s}{\psi}=\{\psi_{i_1\cdots i_s}|i_1,\cdots,i_s=1,\cdots,n\}$. As in §1, we extended the action of the transformation T to the variables $dx=(dx_1,\cdots,dx_n)$, du, $d\underset{1}{u}$ =

$(du_1,\cdots,du_n),\cdots,\ \underset{k}{du}=\{du_{i_1\cdots i_k}|i_1,\cdots,i_k=1,\cdots,n\}$ by means of

$$dx'^i=\frac{\partial f^i}{\partial x^j}dx^j+\frac{\partial f^i}{\partial u}du+\frac{\partial f^i}{\partial u_j}du_j+\cdots+\frac{\partial f^i}{\partial u_{j_1\cdots j_k}}du_{j_1\cdots j_k},$$

$$du'=\frac{\partial \phi}{\partial x^j}dx^j+\frac{\partial \phi}{\partial u}du+\frac{\partial \phi}{\partial u_j}du_j+\cdots+\frac{\partial \phi}{\partial u_{j_1\cdots j_k}}du_{j_1\cdots j_k},$$

(2.2)

$$du_i'=\frac{\partial \psi_i}{\partial x^j}dx^j+\frac{\partial \psi_i}{\partial u}du+\frac{\partial \psi_i}{\partial u_j}du_j+\cdots+\frac{\partial \psi_i}{\partial u_{j_1\cdots j_k}}du_{j_1\cdots j_k},$$

$$du'_{i_1\cdots i_k}=\frac{\partial \psi_{i_1\cdots i_k}}{\partial x^j}dx^j+\frac{\partial \psi_{i_1\cdots i_k}}{\partial u}du+\frac{\partial \psi_{i_1\cdots i_k}}{\partial u_j}du_j+\cdots+\frac{\partial \psi_{i_1\cdots i_k}}{\partial u_{j_1\cdots j_k}}du_{j_1\cdots j_k}.$$

The combined action of (2.1) and (2.2) we call the prolonged transformation $\tilde{T}$.

DEFINITION 2.1. The transformation T is called a *kth-order tangent transformation* if the kth-order tangency conditions

(2.3)

$$du-u_j\,dx^j=0,$$
$$du_{i_1\cdots i_s}-u_{i_1\cdots i_s j}\,dx^j=0,\qquad s=1\cdots k-1$$

are invariant with respect to the action of the prolonged transformation $\tilde{T}$.

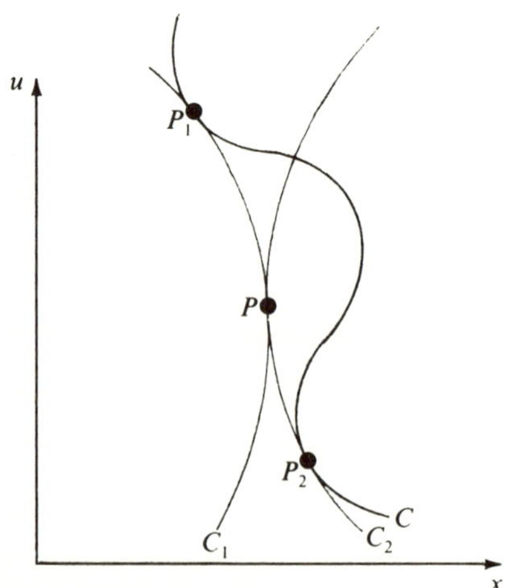

FIG. 3. The curve C_1 is first-order tangent to the curve C_2 at the point P, and the curve C osculates with C_1, C_2 at the points P_1, P_2, respectively.

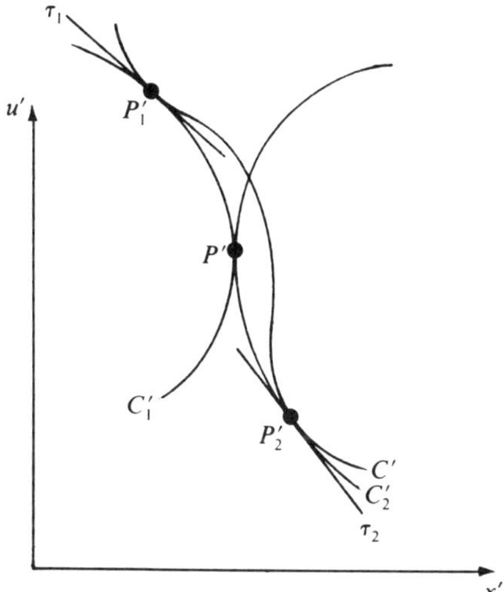

FIG. 4. The image of the geometrical configuration shown in Fig. 3 under an osculating transformation where in addition τ_1, τ_2 are tangent to C_1', C_2' at the points P_1', P_2', respectively.

Bäcklund [1], [2] proved, again under the assumption that (2.1) is an invertible map, that there are no kth-order tangent transformations beyond Lie tangent transformations extended to the variables $\underset{2}{u},\cdots,\underset{k}{u}$ through differentiation. Here we present a more detailed version of Bäcklund's original elegant geometrical proof of this fact for second-order tangent transformations involving one "independent" variable x [i.e., in (2.1) take $k=2$, $n=1$].

The geometrical meaning of a first-order tangent transformation is that transformation (1.1) converts any two curves in (x,u)-space which have first-order tangency at some point into two curves in (x',u')-space which have first-order tangency at the image of the original point of tangency. Similarly, a second-order tangent transformation (or osculating transformation) converts any two curves in osculation (possessing second-order tangency) into two other curves in osculation. Now Bäcklund's result on the nonexistence of invertible second-order tangent transformations beyond those which are extensions of Lie tangent transformations possesses the following geometrical formulation:

THEOREM 2.1 (Bäcklund [1]). *Any invertible sufficiently smooth osculating transformation T is a Lie tangent transformation.*

Proof. Given any two sufficiently smooth curves C_1 and C_2 which are first-order tangent at a point P (Fig. 3), consider the action of T on a neighborhood of the point P. Let $C_i' = T(C_i)$, $i=1$, 2, and $P' = T(P)$ (Fig. 4). Then to prove the theorem, it is necessary to show that the curves C_1' and C_2' have first-order tangency at the point P'. To show this, consider two arbitrary points $P_1 \in C_1$, $P_2 \in C_2$ and a third curve C which osculates with the curves C_1

and C_2 at the points P_1 and P_2, respectively. Let $P_i'=T(P_i)$, $i=1, 2$, and $C'=T(C)$. Because T is an osculating transformation, the curve C' is in osculation with the curves C_1', C_2' at the points P_1', P_2', respectively. Let τ_1, τ_2 be the corresponding tangents at the points P_1', P_2', respectively. Because the points P_1, P_2 are arbitrary, the points P_1', P_2' can be taken arbitrarily near the point P'. Therefore, according to the assumption that T is a sufficiently smooth map and C is a twice differentiable curve, the tangents τ_1 and τ_2 will be arbitrarily close to each other and hence to the tangent at P'. ∎

Remark. According to Definition 2.1, an osculating transformation maps the set S_2 of all pairs of curves (surfaces) which are in osculation into itself. Thus, these transformations do not possess *a priori* the property of acting invariantly on the set $S_1 \supset S_2$ of all curves (surfaces) which have first-order tangency with respect to each other. But Theorem 2.1 establishes a remarkable property of osculating transformations, namely, they leave invariant the set S_1 when it is only required that they leave invariant the subset S_2 of S_1.

One can prove by induction (Bäcklund [2]) a similar statement about the nonexistence of third-order or higher finite-order tangent transformations beyond those of Lie, for an arbitrary number of variables $x^1,\cdots,x^n$. So the following general statement summarizes Bäcklund's results on invertible (single-valued) higher-order tangent transformations.

THEOREM 2.2 (Bäcklund [1], [2]). *Any invertible kth-order tangent transformation* (2.1) *is a prolongation to kth-order derivatives of a Lie tangent transformation* (1.1).

§3. Infinite-Order Structure

Given the variables $x=(x^1,\cdots,x^n)\in\mathbb{R}^n$, $u\in\mathbb{R}$,

$$\underset{s}{u}=\left\{u_{i_1\cdots i_s}\,|\,i_1,\cdots,i_s=1,\cdots,n\right\},\qquad s=1,2,\cdots,$$

consider a transformation T:

$$\begin{aligned}
&x'=f\left(x,u,\underset{1}{u},\cdots\right),\\
&u'=\phi\left(x,u,\underset{1}{u},\cdots\right),\\
&\underset{1}{u}{}'=\underset{1}{\psi}\left(x,u,\underset{1}{u},\cdots\right),\\
&\cdots\cdots\cdots\cdots\cdots
\end{aligned}\tag{3.1}$$

In (3.1), the number of arguments of each of the functions f^i, ϕ is *a priori* arbitrary and may be finite or infinite. The number of equations in (3.1) is assumed to be infinite. Again, the action (3.1) is extended formally to the variables dx, $du, \underset{1}{du}, \cdots$ by means of the following transformation law:

$$
\begin{aligned}
dx'^i &= \frac{\partial f^i}{\partial x^j} dx^j + \frac{\partial f^i}{\partial u} du + \frac{\partial f^i}{\partial u_j} du_j + \cdots, \\
du' &= \frac{\partial \phi}{\partial x^j} dx^j + \frac{\partial \phi}{\partial u} du + \frac{\partial \phi}{\partial u_j} du_j + \cdots, \\
du'_i &= \frac{\partial \psi_i}{\partial x^j} dx^j + \frac{\partial \psi_i}{\partial u} du + \frac{\partial \psi_i}{\partial u_j} du_j + \cdots, \\
&\cdots\cdots\cdots\cdots\cdots\cdots
\end{aligned} \tag{3.2}
$$

where the infinite-dimensional version of the notation defined in §2 is employed here. The transformations (3.1) and (3.2) represent the prolonged transformation $\tilde{T}$. Now we introduce the following definition.

DEFINITION 3.1. The transformation T is called *a Lie–Bäcklund tangent transformation* (∞-order tangent transformation) if the ∞-order tangency conditions

$$du - u_j dx^j = 0, \qquad du_i - u_{ij} dx^j = 0, \qquad \cdots, \tag{3.3}$$

are invariant with respect to the action of the prolonged transformation $\tilde{T}$.

There are transformations (3.1) which are Lie–Bäcklund tangent transformations but which are not simple prolongations of a Lie tangent transformation. For example, Bäcklund considered transformations of the type

$$
\begin{aligned}
x' &= f\left(x, u, \underset{1}{u}, \cdots, \underset{k}{u}\right), \\
u' &= \phi\left(x, u, \underset{1}{u}, \cdots, \underset{k}{u}\right),
\end{aligned} \tag{3.4}
$$

together with their extension to the form (3.1) through differentiation and elimination. Of course, the form (3.4) does not realize the full potential contained in the notion of Lie–Bäcklund tangent transformations as defined in Definition 3.1. For examples see §12.

Bäcklund [2] considered transformations of the form (3.4) for arbitrary $k > 1$ and investigated the following important question: Are there among the ∞-order tangent transformations obtained from (3.4) as described above, those which are closed in a finite-dimensional space, i.e., is it possible to find transformations of the form (3.4) such that after extension to kth-order derivatives, one obtains an invertible map of the closed form (2.1) in $(x, u, \underset{1}{u}, \cdots, \underset{k}{u})$-space? In other words, he explored the question of the existence of a nontrivial generalization of Lie tangent transformations by Lie-Bäcklund transformations under the additional requirement that the latter leave the finite-dimensional space of variables $x, u, \underset{1}{u}, \cdots, \underset{k}{u}$ invariant. If this question could be answered in the affirmative, one would have a Lie–Bäcklund transformation which was not a prolonged Lie tangent transformation, but which converted any kth-order partial differential equation into another one (without raising the order). But Bäcklund found, as Lie [1] expected, that the answer to this question must be negative, and the corresponding results can be formulated as the following theorem.

THEOREM 3.1. *Given a transformation*

$$x'=f\left(x, u, \underset{1}{u}, \cdots, \underset{k}{u}\right),$$
$$u'=\phi\left(x, u, \underset{1}{u}, \cdots, \underset{k}{u}\right), \tag{3.5}$$

the extension of this transformation to derivatives up to the kth order yields a transformation of the closed form

$$x'=f\left(x, u, \underset{1}{u} \ldots \underset{k}{u}\right),$$
$$u'=\phi\left(x, u, \underset{1}{u}, \cdots, \underset{k}{u}\right),$$
$$\cdots\cdots\cdots\cdots\cdots \tag{3.6}$$
$$\underset{k}{u}'=\underset{k}{\psi}\left(x, u, \underset{1}{u}, \cdots, \underset{k}{u}\right),$$

if and only if $k=1$ *and* (3.6) *is a Lie tangent transformation.*

II. TRANSFORMATION OF FAMILIES OF SURFACES

§4. Lie's Second Question

As discussed in §§2, 3, there are no nontrivial generalizations of Lie tangent transformations if one understands a transformation as an invertible map in $\left(x, u, \underset{1}{u}, \cdots, \underset{k}{u}\right)$-space for any finite $k \geqq 1$, or in classical terminology as a single-valued surface transformation. Lie [3] and Bäcklund [3] showed in later papers that one way to realize the sought-after substantive generalization is to consider many-valued surface transformations.

In order to understand the nature of many-valued surface transformations, consider the case of two independent variables and one dependent variable. Here we employ the classical notation x, y, z, p, q, r, t, s for two independent variables and one dependent variable together with its first and second-order partial derivatives. The capital letters will denote the corresponding transformed quantities. Consider a transformation

$$X = F_1(x, y, z, p, q),$$
$$Y = F_2(x, y, z, p, q), \tag{4.1}$$
$$Z = F_3(x, y, z, p, q).$$

For general transformations of this type the extension to first derivatives through differentiation and elimination yields transformation laws for the first derivatives depending not only on x, y, z, p, q, but also on second derivatives, i.e.,

$$P = \psi_1(x, y, z, p, q, r, s, t),$$
$$Q = \psi_2(x, y, z, p, q, r, s, t). \tag{4.2}$$

If r, s, t do not appear in (4.2), we may revert to the case of Lie tangent transformations. In the general case, however, the formulas (4.1) and (4.2) convert any surface in (x, y, z)-space into one surface in (X, Y, Z)-space, but to one surface in (X, Y, Z)-space there corresponds an infinite family of surfaces in (x, y, z)-space. In particular, in the latter case the corresponding family is given by a first-order partial differential equation. For example, this correspondence can be established in the following way. Given a surface

$$Z = H(X, Y), \tag{4.3}$$

the substitution of X, Y, Z given by (4.1) into (4.3) yields a first-order partial differential equation

$$h(x, y, z, p, q) = 0, \tag{4.4}$$

where

$$h(x, y, z, p, q) = F_3(x, y, z, p, q) - H(F_1(x, y, z, p, q), F_2(x, y, z, p, q)). \tag{4.5}$$

Therefore the transformation (4.1) converts one surface given by (4.3) in (X, Y, Z)-space into the family of integral surfaces of the first-order partial differential equation (4.4).

Although in general the transformation (4.1) is not single-valued, it can be used effectively in the transformation theory of differential equations, as was demonstrated by Lie and Bäcklund. In this regard, there are two natural requirements which are imposed upon the transformation (4.1): It must (i) transform a given differential equation into itself or another differential equation of the same or lower order, and (ii) be a surface transformation on a given family of surfaces (i.e., it is required to transform any surface solution of a given differential equation into some surface, which is not necessarily described as a solution of the original equation). In this context, it is the consideration of second or higher-order equations which is of primary interest, because the transformation theory of first-order equations is completely treated by Lie's theory of first-order tangent transformations as applied to these equations. More precisely, requirement (ii) means that if ω is a given differential equation, say a second-order differential equation, then the quantities P, Q given by (4.2) are required to satisfy, for the surface element (x, y, z, p, q) of any surface solution $z=z(x, y)$ of the equation ω, the integrability condition

$$\frac{\partial P}{\partial Y}-\frac{\partial Q}{\partial X}=0 \quad \text{on } \omega. \tag{4.6}$$

The importance of the existence of such transformations was emphasized by Lie in his second question (Lie [1], p. 223).

Lie's Second Question Given a higher-order partial differential equation, does this equation admit a transformation which is not a first-order tangent transformation?

Lie predicted an affirmative answer to this question. Further, he commented that the realization of this possibility would open an important area of investigation. In the next section we present Lie's realization of this possibility through his analytical treatment of Bianchi's construction for surfaces of constant curvature, as well as several of its classical generalizations.

§5. Bianchi–Lie Transformation

Lie's analytical treatment of Bianchi's geometrical construction of a transformation of surfaces of constant curvature was the example which first clearly demonstrated the potential inherent in the notion of many-valued surface transformations. Because this example illustrates the basic notions and techniques for the application of this type of transformation to differential equations, we trace its geometrical origin as well as its analytical expression. This example also directly leads to what are called in the literature Bäcklund transformations and clarifies their structure.

Here we outline Bianchi's geometrical construction. In three-dimensional Euclidean space, consider a surface $\mathbb{S}$ of constant negative curvature $-1/a^2$, where the constant $a>0$, and another surface $\mathbb{S}'$ which is related to $\mathbb{S}$ in the following way (see Fig. 5). To every point $\mathrm{M}\in\mathbb{S}$ there corresponds a point $\mathrm{M}'\in\mathbb{S}'$ such that:

(i) $|\mathrm{MM}'|=a$, where $|\mathrm{MM}'|$ is the length of the line segment MM'; if τ, τ' are tangent planes to $\mathbb{S}, \mathbb{S}'$ at M, M′ respectively, then

(ii) $\mathrm{MM}'\in\tau$,

(iii) $\mathrm{MM}'\in\tau'$,

(iv) $\tau\perp\tau'$.

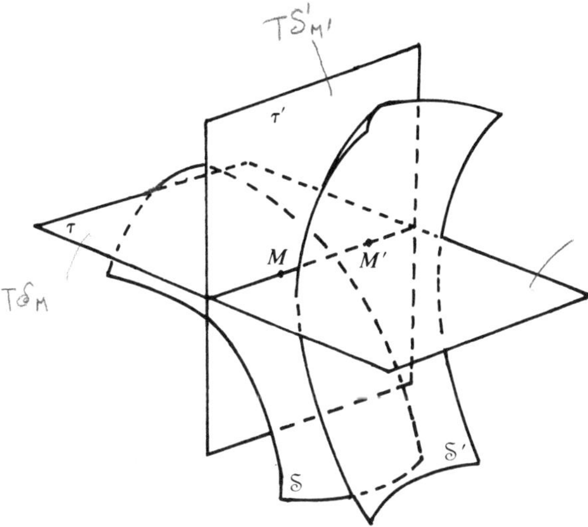

FIG. 5. Bianchi's geometrical construction for surfaces of constant negative curvature.

Bianchi [1] demonstrated that $\mathbb{S}'$ is also a surface of the same constant curvature $-1/a^2$.

In order to clarify the analytical structure of this transformation and investigate the properties of the family of transformed surfaces $\mathbb{S}'$, Lie expressed Bianchi's geometrical construction in an equivalent analytical form. Any surface $\mathbb{S}$ of constant curvature $-1/a^2$ satisfies the second-order partial differential equation

$$s^2 - rt = \frac{1+p^2+q^2}{a^2}, \tag{5.1}$$

where the usual notation p, q, r, s, t is employed for the first and second derivatives. If (X, Y, Z, P, Q) is a surface element of the transformed surface $\mathbb{S}'$, then conditions (i)–(iv) in Bianchi's construction can be expressed in the following form, which we call the Bianchi–Lie transformation:

$$\begin{aligned} (x-X)^2+(y-Y)^2+(z-Z)^2&=a^2,\\ p(x-X)+q(y-Y)-(z-Z)&=0,\\ P(x-X)+Q(y-Y)-(z-Z)&=0,\\ pP+qQ+1&=0. \end{aligned} \tag{5.2}$$

First we observe that, given any surface element (x, y, z, p, q), equation (5.2) gives four relationships between the five quantities X, Y, Z, P, Q; hence there is a onefold infinity of surface elements (X, Y, Z, P, Q) satisfying (5.2). It then follows that since (5.2) represents an ∞-valued surface transformation, it cannot describe a Lie tangent transformation (see Definition 1.1 and §4).

In order to treat (5.2) as a surface-transformation, consider z as a given function of x, y and regard p, q, r, s, t as the usual first- and second-order derivatives of this function. Let $\mathbb{S}$ denote the surface described by the function $z=z(x,y)$. Then we seek the condition for any element (X, Y, Z, P, Q) obtained by means of (5.2) from a surface element (x, y, z, p, q) of the surface $\mathbb{S}$ to also be a surface element, i.e.,

$$\frac{\partial P}{\partial Y}-\frac{\partial Q}{\partial X}=0 \quad \text{on } \mathbb{S}. \tag{5.3}$$

Lie established this condition as a consequence of the following result, which is formulated here as

LEMMA 5.1. *Given a surface element* (x, y, z, p, q) *of a surface* $\mathbb{S}$, *suppose that* (x, y, z, p, q), (X, Y, Z, P, Q) *are related by* (5.2). *If* (X, Y, Z, P, Q) *is a surface element, i.e., if*

$$\frac{\partial P}{\partial Y}=\frac{\partial Q}{\partial X} \quad \text{on } \mathbb{S}, \tag{5.4}$$

then $\mathbb{S}$ *is a surface of constant curvature, i.e.,* $\mathbb{S}$ *satisfies* (5.1).

Outline of Lie's proof. Differentiating the first two relations in (5.2) and taking into account that $dz=p\,dx+q\,dy$, $dZ=P\,dX+Q\,dY$, one obtains expressions for dx, dy as linear functions of dX, dY with variable coefficients. By applying similar operations to the last two relations in (5.2) and taking into account the previous result for dx, dy and the equations $dp=r\,dx+s\,dy$, $dq=s\,dx+t\,dy$, one obtains expressions for dP, dQ as linear functions of dX, dY. From these expressions one finally obtains

$$\sigma\left(\frac{\partial P}{\partial Y}-\frac{\partial Q}{\partial X}\right)=s^2-rt-\frac{1+p^2+q^2}{a^2}, \tag{5.5}$$

where σ is a specified function of $(x, y, z, p, q, r, s, t, X, Y, Z, P, Q)$. Therefore, if (5.4) applies, then $\mathbb{S}$ satisfies (5.1), i.e., $\mathbb{S}$ is a surface of constant curvature.

Note 5.1. A statement dual to Lemma 5.1 holds because (5.2) is symmetric with respect to the interchange of the elements (x, y, z, p, q) and (X, Y, Z, P, Q).

As a direct consequence of Lemma 5.1 and Note 5.1 one obtains

LEMMA 5.2. *The differential equation* (5.1) *is invariant under the transformation* (5.2) *in the following sense. Suppose* $\mathbb{S}$ *is a surface of constant curvature* $-1/a^2$ *and* $\mathbb{S}'$ *is an image of* $\mathbb{S}$ *under the action of* (5.2); *then* $\mathbb{S}'$ *is also a surface of the same constant curvature* $-1/a^2$.

Summarizing these results, one obtains the following theorem, due to Lie, for the transformation (5.2) of surfaces of constant curvature.

THEOREM 5.1 (Lie [3]). *Equation* (5.1) *admits the transformation* (5.2), *and the transformation* (5.2) *is defined only on solutions of* (5.1).

According to Theorem 5.1 and the structure of the transformation (5.2), one can construct by quadratures a family of surfaces of constant curvature starting from some given one. This possibility (i.e., of employing many-valued generalizations of Lie tangent transformations for integrating second-order partial differential equations) was first realized by Bianchi [1] for a special type of surface of constant curvature; it was developed by Lie [3], [4] and, as stated before, recognized by him as opening up a new area of investigation. More precisely, with this example Lie demonstrated that many-valued surface transformations can be used in a manner similar to that in his previous theory of differential equations based on first-order tangent and point transformations. In particular, (5.2), regarded as a transformation, is admitted by the differential equation (5.1), and as a direct consequence of the definition of invariance one has the property that such a transformation converts a solution of the equation (5.1) into a family of solutions of the same equation. For a given solution $z=z(x,y)$ of (5.1), the corresponding transformed family of surfaces is found by quadratures, namely, here as a solution of a completely integrable system of two first-order partial differential equations for the unknown function $Z(X,Y)$, which is obtained from (5.2) when one substitutes $z=z(x,y)$ into (5.2) and then uses two of these equations to eliminate the variables x and y. This procedure, applied to four general equations relating $(x, y, z, p, q),(X, Y, Z, P, Q)$ instead of (5.1), leads to transformations which are called Bäcklund transformations in the literature.

§6. Bäcklund Transformations

The results of Bianchi and Lie in their geometrical and analytical treatments, respectively, of the transformation properties of surfaces of constant curvature led investigators to search for generalizations of the Bianchi–Lie many-valued surface transformation discussed in §5. Here we distinguish different types of these generalizations, some of which have been separately identified in the classical literature as Bäcklund transformations. The first basic type discussed in this section is based on a geometrical generalization of Bianchi's construction. In presenting this discussion we will follow Darboux's complete presentation of this first generalization. The second basic type, which is based on Lie's analytical treatment of Bianchi's operation, actually subsumes many subtypes, all of which are connected with the consistency of an overdetermined system of first-order partial differential equations. Finally, we shall relate these subtypes to the types previously discussed in §3 which we have termed a Lie–Bäcklund tangent transformation.

Bäcklund [5] generalized Bianchi's result (see §5) to surfaces $\mathbb{S}$, $\mathbb{S}'$ that are related by a modification of Bianchi's construction which is obtained by replacing the condition of orthogonality of the two tangent planes τ, τ' with the condition that the angle between these tangent planes is fixed, i.e., instead of Bianchi's condition (iv), Bäcklund's condition is

(iv′) $\angle(\tau, \tau')=\text{const.}$

The Bianchi–Lie transformation (5.2) is then replaced by the following equations:

$$
\begin{aligned}
(x-X)^2+(y-Y)^2+(z-Z)^2&=a^2,\\
p(x-X)+q(y-Y)-(z-Z)&=0,\\
P(x-X)+Q(y-Y)-(z-Z)&=0,\\
pP+qQ+1-b\sqrt{1+p^2+q^2}\,\sqrt{1+P^2+Q^2}&=0.
\end{aligned}
\tag{6.1}
$$

The geometrical significance of the constant b will be explained in connection with Darboux's more general considerations. Bäcklund proved that (6.1) is a surface transformation only for a surface $\mathbb{S}$ of constant negative curvature $-1/a^2$ and that $\mathbb{S}'$ is a surface of the same constant curvature. This is the analogue of Lie's Theorem 5.1. In the classical geometrical literature, the transformation (6.1) of surfaces of constant curvature is called a Bäcklund transformation.

Darboux improved the presentation of this result (Darboux [1], pp. 442–444). In particular, Darboux completed the geometrical analysis of Bianchi's construction by first replacing Bianchi's conditions (ii), (iii) with the conditions that the line segment MM′ makes fixed angles (not necessarily equal to each other) with the tangent planes τ, τ'. However, he retained Bäcklund's condition (iv′). These considerations lead to the equations

$$
\begin{aligned}
(x-X)^2+(y-Y)^2+(z-Z)^2&=a^2,\\
p(x-X)+q(y-Y)-(z-Z)+ba\sqrt{1+p^2+q^2}&=0,\\
P(x-X)+Q(y-Y)-(z-Z)+b'a\sqrt{1+P^2+Q^2}&=0,\\
pP+qQ+1-c\sqrt{1+p^2+q^2}\,\sqrt{1+P^2+Q^2}&=0.
\end{aligned}
\tag{6.2}
$$

Here b,b',c are constants with the following geometrical meaning: b, b' are the trigonometric sines of the angles between MM′ and τ, τ', respectively; c is the cosine of the angle between τ, τ'. Then he classified the cases for which (6.2) is a surface transformation. In particular, he recovered Bäcklund's generalization of the Bianchi–Lie transformation, namely, for the case that $c^2-1\neq 0$ and $b=b'=0$, equation (6.2) becomes (6.1).

Further, Bäcklund's geometrical generalization turned out to be analytically equivalent to a Bianchi–Lie transformation up to a one-parameter group of dilatations. To clarify this connection, we present the classical discussion (Darboux [1], Chapter 12; Bianchi [2], §262) of surfaces of constant curvature in terms of the sine–Gordon equation

$$
\frac{\partial^2}{\partial\alpha\,\partial\beta}(2\theta)=\sin 2\theta.
\tag{6.3}
$$

In this representation the Bianchi–Lie transformation of solutions of (6.3) is given in the form

$$\frac{\partial(\theta+\theta')}{\partial\alpha}=\sin(\theta-\theta'),\qquad \frac{\partial(\theta-\theta')}{\partial\beta}=\sin(\theta+\theta'), \tag{6.4}$$

while Bäcklund's generalization has the form

$$\frac{\partial(\theta+\theta')}{\partial\alpha}=\frac{1}{a}\sin(\theta-\theta'),\qquad \frac{\partial(\theta-\theta')}{\partial\beta}=a\sin(\theta+\theta'), \tag{6.5}$$

with an arbitrary constant $a\neq 0$. Now using the invariance of (6.3) with respect to the group of dilatations

$$\alpha'=\alpha a,\qquad \beta'=\beta/a, \tag{6.6}$$

one can directly transform (6.5) into the canonical form (6.4).

Thus, although Bäcklund's construction [called in the classical geometrical literature (Darboux [1], Bianchi [2]) a Bäcklund transformation] geometrically generalized Bianchi's construction, analytically it is only a simple composition of the Bianchi–Lie equation (6.4) and a one-parameter Lie group of dilatations (6.6). A nontrivial analytical generalization of the Bianchi–Lie transformation was introduced by Bäcklund [3], who considered four general relations between two sets of surface elements:

$$F_i(x,y,z,p,q,X,Y,Z,P,Q)=0,\qquad i=1,2,3,4. \tag{6.7}$$

A literal repetition of Lie's considerations and techniques for treating (5.2) applied to (6.7) leads to what is called in the literature a Bäcklund transformation—in the analytical sense, as contrasted with the prior geometrical sense (Clairin [1], Goursat [2], Ames [1]). In particular, we recall that given (5.2), Lie posed and solved the problem of determining the family of surfaces in (x, y, z)-space such that (5.2) acts on a member of this family as a surface transformation in the previously described sense. His result, as expressed by Theorem 5.1, is that this family is a family of surfaces of constant curvature. Now turning to (6.7), and following Lie, we see that substituting a given function $z=z(x,y)$ in (6.7), and using two of the resulting relations to eliminate x, y, the two remaining relationships implied by (6.7) represent an overdetermined system of two first-order partial differential equations in one unknown function, which we denote as

$$f(X,Y,Z,P,Q)=0,\qquad \phi(X,Y,Z,P,Q)=0. \tag{6.8}$$

The consistency conditions for this system have the form of partial differential equations for the function $z(x,y)$ and are stated in their general form in Bäcklund's paper (Bäcklund [3], p. 311). If $z(x, y)$ satisfies these consistency conditions, then (6.7) is considered as a transformation of the surface $z=z(x, y)$ in (x, y, z)-space into a surface in (X, Y, Z)-space, and this later surface is given as a solution surface of the system (6.8), now treated as an integrable system.

Lie's treatment of Bianchi's construction when extended to (6.7) leads to the consideration of several types of surface transformations, all christened Bäcklund transformations in the literature. The variety of types of these transformations is connected with the possibility of posing different problems when one treats (6.7) as a surface transformation. The sources of these possibilities lie principally in applications.

The most extensively discussed type in the classical literature is the one defined by second-order partial differential equations in the following sense. Equation (6.7) has the property that elimination of (X, Y, Z, P, Q) reduces it to a second-order partial differential equation ω for $z(x, y)$, and conversely elimination of (x, y, z, p, q) reduces it to a second-order partial differential equation Ω for $Z(X, Y)$.

Lie's original problem of the determination of invariance transformations for a given system of differential equations applied to (6.7) gives rise to another type of Bäcklund transformation, and there are many possibilities for further generalization of these types of transformations (Goursat [2]).

Analyzing what is common to all these types of transformations, one can define a Bäcklund transformation as one that is specified by an overdetermined system of first-order differential equations of the form (6.7). As Goursat [2] has remarked, one can generalize this form in many ways, including increasing the dimension of the underlying space(s), the order of the surface elements, and the number of relations in (6.7), etc.

We conclude this section with the remark that the Bäcklund transformations of the form (6.7) can be related to Lie–Bäcklund tangent transformations as defined in §3 in the following way. Here we confine ourselves to a heuristic argument which is based on the observation (e.g. Bäcklund [1]) that a surface $z=z(x, y)$ in (x, y, z)-space is specified at any fixed (x, y) in point language by the values of z and its derivatives $p, q, r, s, t, \ldots$ at (x, y). Without loss of generality we take

$$\begin{aligned} X&=x,\\ Y&=y,\\ P&=\psi_1(x, y, z, p, q, Z),\\ Q&=\psi_2(x, y, z, p, q, Z) \end{aligned} \tag{6.9}$$

to represent a given Bäcklund transformation of a family of surfaces M. Now if the surface $z=h(x, y)$ belongs to the family M, the equations (6.9) can be integrated. Let $Z=Z_h(x, y)$ be a particular solution for a given h. Then, if we take $z=h(x,y)$ to be an arbitrary element of M and specify any surface by the element $(x, y, z, p, q, r, s, t, \ldots)$, e.g., this can be obtained via a Taylor series representation as in Examples 3 and 4 of §12, the solution $Z_h(x, y)$ becomes $Z=\phi(x, y, z, p, q, r, s, t, \ldots)$. This formula, when added to (6.9) and extended by differentiation and elimination, yields a transformation law of the form (3.1):

(6.10)
$$\begin{aligned} &X=x, \qquad Y=y,\\ &Z=\phi\,(x, y, z, p, q, r, s, t, \ldots),\\ &P=\psi_1(x, y, z, p, q, \phi(x, y, z, p, q, r, s, t, \ldots)),\\ &Q=\psi_2(x, y, z, p, q, \phi(x, y, z, p, q, r, s, t, \ldots))\\ &\ldots\ldots\ldots\ldots\ldots\ldots\ldots\ldots \end{aligned}$$

III. EXAMPLES OF BÄCKLUND TRANSFORMATIONS

§7. Invariance Transformations

Here we shall consider two Bäcklund transformations of invariant type—that is, both dependent variables z and z_1 satisfy the same equation. The first example concerns the classical sine–Gordon equation $z_{xy}=\sin z$, and the second the Korteweg–de Vries equation $u_y+6uu_x+u_{xxx}=0$.

In what follows we proceed from (6.9) rewritten as

(7.1) $$p=f_1(x, y, z, z_1, p_1, q_1), \qquad q=f_2(x, y, z, z_1, p_1, q_1),$$

where $p=z_x$, $q=z_y$, $p_1=(z_1)_x$, $q_1=(z_1)_y$, and supplement it with the integrability condition $dp/dy=dq/dx$. This condition generates the relationship

(7.2)
$$\begin{aligned} &\frac{\partial f_1}{\partial y}+f_2\frac{\partial f_1}{\partial z}+q_1\frac{\partial f_1}{\partial z_1}+s_1\frac{\partial f_1}{\partial p_1}+t_1\frac{\partial f_1}{\partial q_1}\\ &\qquad=\frac{\partial f_2}{\partial x}+f_1\frac{\partial f_2}{\partial z}+p_1\frac{\partial f_2}{\partial z_1}+r_1\frac{\partial f_2}{\partial p_1}+s_1\frac{\partial f_2}{\partial q_1}, \end{aligned}$$

where r_1, s_1, and t_1 represent $(z_1)_{xx}$, $(z_1)_{xy}$, and $(z_1)_{yy}$, respectively. Equation (7.2) is linear in r_1, s_1, and t_1, and in general depends upon x, y, z, z_1, p_1, and q_1.

Suppose our initial concern is with the two simultaneous equations (7.1). When z occurs in (7.2), we can think of that equation as solved so as to express z in terms of x, y, z_1, p_1, q_1, r_1, s_1, and t_1. When the value of z so obtained is substituted into the first-order equations (7.1), they become two equations of the third order for the determination of z_1. From general theory it is known that unless the original equations cannot be solved with respect to p and p_1, or with respect to q and q_1, they possess common integrals. Consequently, the two third-order equations which are satisfied by z_1 must be compatible. They must therefore lead to values of z_1 that involve arbitrary functions. If the original equations (7.1) are of second order, then the equations for z_1 will be of fourth order in the preceding argument.

If the integrability condition (7.2) is free of z, then it becomes a single second-order equation for z_1. Upon solving this, the z_1 so obtained is substituted into (7.1), and a quadrature of those equations leads to a value of z containing an arbitrary constant. An exceptional case arises when (7.2) does not contain r_1, s_1, and t_1—i.e., when

$$\frac{\partial f_1}{\partial q_1}=0, \qquad \frac{\partial f_2}{\partial p_1}=0, \qquad \frac{\partial f_1}{\partial p_1}=\frac{\partial f_2}{\partial q_1}.$$

Here, if z is involved, then z_1 satisfies two equations of the second order. If z is not present, then z_1 satisfies a single equation of the first order.

We proceed now to give the detailed construction of two Bäcklund invariance transformations. As will be seen, this is primarily a study of overdetermined systems. Many arbitrary functions will appear in the analysis, and educated choices of them simplify the analysis while still leading to the desired results. The transformations are *not* unique.

The Sine-Gordon Equation. A detailed discussion of the sine–Gordon equation

$$(z_1)_{xy}=\sin z_1 \tag{7.3}$$

and its many applications is given in Barone et al.[1] We will use (7.3) as a carrier for demonstrating a classical method (Bäcklund [3], Clairin [1])[2] for generating a Bäcklund transformation which leaves a given differential equation invariant.

Consider the symmetric explicit special case of (7.1)

$$p=f\,(p_1, z, z_1), \qquad q=\psi\,(q_1, z, z_1),$$

where z_1 is a solution of (7.3) and z also satisfies the same equation. The integrability requirement $dp/dy=dq/dx$ generates the relation [compare (7.2)]

$$\Omega=\sin z_1\left(\frac{\partial f}{\partial p_1}-\frac{\partial \psi}{\partial q_1}\right)-p_1\frac{\partial \psi}{\partial z_1}+q_1\frac{\partial f}{\partial z_1}+\psi\frac{\partial f}{\partial z}-f\frac{\partial \psi}{\partial z}=0. \tag{7.4}$$

Calculation of successive derivatives of Ω to the point where z_1 is no longer

[1]A. Barone, F. Esposito, C. J. Magee, and A. C. Scott, *Theory and applications of the sine-Gordon equation*, Riv. Nuovo Cimento (2), 1(1971), p. 227.

[2]A. R. Forsyth, *Theory of Differential Equations*, vol. 6, 1906 (reprinted by Dover, New York, 1959), Chapter 21.

explicitly present yields the following equations:

$$\text{(7.5)}\qquad \frac{\partial\Omega}{\partial p_1}=\sin z_1\frac{\partial^2 f}{\partial p_1^2}-\frac{\partial\psi}{\partial z_1}+q_1\frac{\partial^2 f}{\partial p_1\partial z_1}+\psi\frac{\partial^2 f}{\partial p_1\partial z}-\frac{\partial f}{\partial p_1}\frac{\partial\psi}{\partial z}=0,$$

$$\text{(7.6)}\qquad \frac{\partial\Omega}{\partial q_1}=-\sin z_1\frac{\partial^2\psi}{\partial q_1^2}-p_1\frac{\partial^2\psi}{\partial q_1\partial z_1}+\frac{\partial f}{\partial z_1}+\frac{\partial\psi}{\partial q_1}\frac{\partial f}{\partial z}-f\frac{\partial^2\psi}{\partial q_1\partial z}=0,$$

$$\text{(7.7)}\qquad \frac{\partial^2\Omega}{\partial p_1\partial q_1}=-\frac{\partial^2\psi}{\partial q_1\partial z_1}+\frac{\partial^2 f}{\partial p_1\partial z_1}+\frac{\partial\psi}{\partial q_1}\frac{\partial^2 f}{\partial p_1\partial z}-\frac{\partial f}{\partial p_1}\frac{\partial^2\psi}{\partial q_1\partial z}=0,$$

$$\text{(7.8)}\qquad \frac{\partial^3\Omega}{\partial p_1^2\partial q_1}=\frac{\partial^3 f}{\partial p_1^2\partial z_1}+\frac{\partial\psi}{\partial q_1}\frac{\partial^3 f}{\partial p_1^2\partial z}-\frac{\partial^2 f}{\partial p_1^2}\frac{\partial^2\psi}{\partial q_1\partial z}=0,$$

$$\text{(7.9)}\qquad \frac{\partial^3\Omega}{\partial p_1\partial q_1^2}=-\frac{\partial^3\psi}{\partial q_1^2\partial z_1}+\frac{\partial^2\psi}{\partial q_1^2}\frac{\partial^2 f}{\partial p_1\partial z}-\frac{\partial f}{\partial p_1}\frac{\partial^3\psi}{\partial q_1^2\partial z}=0,$$

$$\text{(7.10)}\qquad \frac{\partial^4\Omega}{\partial p_1^2\partial q_1^2}=\frac{\partial^2\psi}{\partial q_1^2}\frac{\partial^3 f}{\partial p_1^2\partial z}-\frac{\partial^2 f}{\partial p_1^2}\frac{\partial^3\psi}{\partial q_1^2\partial z}=0.$$

Equation (7.10) is free of explicit dependence upon z_1, although, of course, solutions of that equation will depend parametrically upon z_1. Upon integration of (7.10), arbitrary functions will appear which are determined by the requirement that (7.5)–(7.9) must also be satisfied. It is possible to separate the p_1 and q_1 dependence of (7.10) and write it as

$$\text{(7.11)}\qquad \frac{\partial^3\psi/\partial q_1^2\partial z}{\partial^2\psi/\partial q_1^2}=\frac{\partial^3 f/\partial p_1^2\partial z}{\partial^2 f/\partial p_1^2}=h(z,z_1).$$

From the usual separation argument it follows that the two left-hand terms of (7.11) do not depend upon p_1 and q_1 (as was expected). Hence the introduction of $h(z, z_1)$. Recalling that $f=f(p_1, z, z_1)$, $\psi=\psi(q_1, z, z_1)$, integration of (7.11) gives

$$\text{(7.12)}\qquad f=g(p_1,z_1)H(z,z_1)+l(z,z_1)p_1+m(z,z_1),$$

$$\text{(7.13)}\qquad \psi=\theta(q_1,z_1)H(z,z_1)+\lambda(z,z_1)q_1+\mu(z,z_1),$$

where g, θ, l, λ, m, μ, and $H=\exp[\int h(z,z_1)dz]$ are arbitrary.

Choosing $H(z,z_1)=1$ simplifies the subsequent analysis, so we shall present only that case. Upon substituting (7.12) and (7.13) into (7.9) we find

$$\text{(7.14)}\qquad \frac{\partial^3\theta}{\partial z_1\partial q_1^2}-\frac{\partial^2\theta}{\partial q_1^2}\frac{\partial l}{\partial z}=0.$$

Consequently $\partial l/\partial z$ is independent of z. Integration of (7.14) generates the solution for θ,

$$\theta(q_1,z_1)=v(q_1)e^{u(z_1)}+r(z_1)q_1+w(z_1), \tag{7.15}$$

where

$$z\frac{du}{dz_1}+\text{const}=l(z,z_1), \tag{7.16}$$

and v, r, and w are arbitrary functions of their arguments. When θ and l are substituted into (7.12) and (7.13), one finds there is no loss in generality in setting $w(z_1)$ and the constant in (7.16) equal to zero.

Corresponding results are obtained from (7.8) for g and λ. At this point, the calculation is further restricted by setting $u=0$ in (7.15) and similarly in the calculation for λ. Thus

$$f(p_1,z,z_1)=\bar{v}(p_1)+\bar{r}(z_1)p_1+m(z,z_1), \tag{7.17}$$

$$\psi(q_1,z,z_1)=v(q_1)+r(z_1)q_1+\mu(z,z_1). \tag{7.18}$$

Next, by substituting (7.17) and (7.18) into (7.7), we find that

$$\frac{d}{dz_1}(r-\bar{r})=0. \tag{7.19}$$

Turning now to (7.6), we find that it imposes the restriction

$$-\sin z_1\frac{d^2v}{dq_1^2}+\frac{\partial m}{\partial z}\frac{dv}{dq_1}=-\frac{\partial m}{\partial z_1}-r(z_1)\frac{\partial m}{\partial z}, \tag{7.20}$$

whose solution will yield the general form of $v(q_1)$. Since the right-hand side depends only upon z and z_1, so must the left-hand side. Thus, unless[3] $\partial m/\partial z=\sin z_1$, one must have $v(q_1)=c_1q_1+c_2$, where c_i are absolute constants. Following up this case, we see that c_1q_1 can be absorbed into the term $r(z_1)q_1$ and c_2 into $\mu(z,z_1)$. Consequently, without loss of generality, we take

$$\psi(q_1,z,z_1)=r(z_1)q_1+\mu(z,z_1) \tag{7.21}$$

and determine m by solving

$$\frac{\partial m}{\partial z_1}+r(z_1)\frac{\partial m}{\partial z}=0, \tag{7.22}$$

which is the simplified form of (7.20). A similar procedure applied to (7.5) leads to

$$f(p_1,z,z_1)=\bar{r}(z_1)p_1+m(z,z_1) \tag{7.23}$$

[3]In this case, we may have $(d^2v/dq_1^2)-(dv/dq_1)=\alpha$, or $v=\bar{A}e^{q_1}-\alpha q_1+\bar{B}$, where α, $\bar{A}$, and $\bar{B}$ are constants.

and the equation for μ,

$$\frac{\partial \mu}{\partial z_1} + \bar{r}(z_1)\frac{\partial \mu}{\partial z} = 0. \tag{7.24}$$

Finally, (7.4) yields

$$(\bar{r} - r)\sin z_1 + \mu\frac{\partial m}{\partial z} - m\frac{\partial \mu}{\partial z} = 0. \tag{7.25}$$

When one sets $\bar{r} = 1$, $r = -1$, equations (7.22) and (7.24) have the simple solutions

$$m = m(z + z_1), \qquad \mu = \mu(z - z_1), \tag{7.26}$$

and (7.25) now becomes

$$2\sin z_1 + \frac{\partial}{\partial z_1}\left[m(z + z_1)\mu(z - z_1)\right] = 0, \tag{7.27}$$

a functional differential equation. Upon differentiation with respect to z, equation (7.27) is expressible as

$$\frac{m''(z + z_1)}{m(z + z_1)} = \frac{\mu''(z - z_1)}{\mu(z - z_1)} = -K^2,$$

where K is a constant and the primes indicate differentiation with respect to the indicated argument. The classical solutions to this system are

$$m(z + z_1) = A\cos K(z + z_1) + B\sin K(z + z_1),$$

$$\mu(z - z_1) = \alpha\cos K(z - z_1) + \beta\sin K(z - z_1).$$

When these are substituted into (7.27) and evaluated at $z = 0$, and odd and even functions of z_1 are equated, one finds

$$\alpha B - \beta A = 0,$$

$$\alpha A + \beta B = 4,$$

$$K = \tfrac{1}{2}.$$

These are satisfied by setting $A = \alpha = 0$, $\beta B = 4$, or $B/2 = 2/\beta = a$. Equations (7.21) and (7.23) now yield *one* of the desired Bäcklund transformations

$$\begin{aligned} \tfrac{1}{2}(q + q_1) &= \frac{1}{a}\sin\tfrac{1}{2}(z - z_1), \\ \tfrac{1}{2}(p - p_1) &= a\sin\tfrac{1}{2}(z + z_1), \end{aligned} \tag{7.28}$$

where both z and z_1 satisfy the sine–Gordon equation (7.3). Hence (7.28) is an invariance transformation for (7.3) in the sense of §5. Of course, in general the condition of invariance is not required. Bäcklund transformations which are not invariance transformations for given differential equations have utility and will be discussed in the next section. In particular, it is even possible to perform a transformation relating a nonlinear to a linear equation.

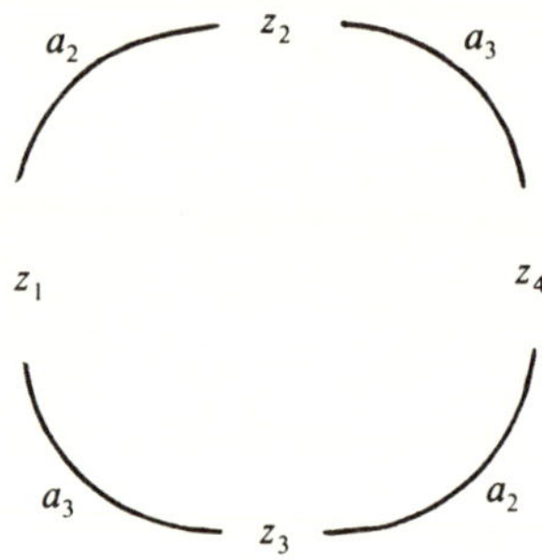

FIG. 6. Diagram for (7.29).

Equation (7.3) was studied many years ago in connection with the theory of pseudospherical surfaces (Darboux [1, p. 432] Bianchi [2]). In Bianchi ([1], [2]) it is shown that a *theorem of permutability* (today called a *nonlinear superposition*—see Chaper 4) exists for this equation. Beginning with a solution z_1 of (7.2), let us generate solutions z_2 and z_3 through a_2 and a_3 in (7.28), respectively; then there is a solution z_4 which is generated from z_2 through a_3 and also from z_3 through a_2. This permutability was expressed by Bianchi ([2], p. 743) in the diagram in Fig. 6. From (7.28) the analytic expressions for these solutions can be written, and then it can be shown that they are related by the expression

$$\tan\left(\frac{z_4-z_1}{4}\right)=\frac{a_2+a_3}{a_2-a_3}\tan\left(\frac{z_2-z_3}{4}\right), \tag{7.29}$$

which is the classical *theorem of permutability*. The knowledge of three solutions enables one to recursively generate an infinite sequence of particular solutions to the sine–Gordon equation.

Perhaps the nonlinear superposition (7.29) should be expected, since each of the equations in (7.28) is transformable into a Riccati equation

$$\frac{dy}{dx}+Q(x)y+R(x)y^2=P(x), \tag{7.30}$$

which is known to have a nonlinear superposition, the so-called *cross ratio*.[4] If y_1, y_2, y_3, and y_4 are particular solutions of (7.30), the cross ratio

$$\frac{(y_1-y_3)(y_2-y_4)}{(y_1-y_4)(y_2-y_3)}=\lambda \tag{7.31}$$

of these solutions is a constant, denoted by λ in (7.31).

Now setting $r=\tan\frac{1}{4}(z+z_1)$ into the second equation of (7.28) results in

$$r_x+ar-\tfrac{1}{2}p\,(1+r^2)=0,$$

[4]H. T. Davis, *Introduction to Nonlinear Differential and Integral Equations* (Dover, 1962).

a Riccati equation. We shall see the same phenomenon occurring in the next section and will elaborate on the idea in Chapter 4.

The Korteweg–de Vries (KdV) Equation. As a second detailed example, an invariance Bäcklund transformation of the KdV equation

(7.32) $$u_y + 6uu_x + u_{xxx} = 0$$

is constructed following Lamb.[5] Upon setting

$$z(x,y) = \int_{-\infty}^{x} u(x',y)\,dx',$$

equation (7.32) becomes

(7.33) $$q + 3p^2 + \alpha = 0$$

after integration and the discard of an arbitrary function of integration. Here $\alpha = u_{xxx}$, while p and q are as before. An invariance Bäcklund transformation of (7.32) can be recovered from that for (7.33) by using the Lie–Bäcklund transformation $z_x = u$.

A form must now be chosen for the two lower-order equations that will constitute an invariance Bäcklund transformation for (7.33). Motivated by the structure of the transformation (7.28) for the sine–Gordon equation, a possible choice is

(7.34a) $$p = f(z, z_1, p_1),$$

(7.34b) $$q = \bar{\phi}(z, z_1, q_1, r, r_1, p, p_1),$$

where both z and z_1 satisfy (7.33). However, from (7.34a), $r = f_z f + f_{z_1 p_1} + f_{p_1 r_1}$, and it is apparent that (7.34b) can be replaced by

(7.34b′) $$q = \phi(z, z_1, q_1, p_1, r_1).$$

The subsequent analysis will involve (7.34a) and (7.34b′).

We now require that z_1 satisfy (7.33) and that $dp/dy = dq/dx$. The second condition generates the relation

(7.35) $$\Omega \equiv (f_{p_1} - \phi_{q_1})s_1 + f_z q + f_{z_1} q_1 - \phi_z p - \phi_{z_1} p_1 + \phi_{r_1}(q_1 + 3p_1^2) = 0,$$

which is the analogue of (7.4). Now

(7.36) $$\Omega_{s_1} = f_{p_1} - \phi_{q_1} = 0,$$

which implies, since f is independent of q_1 and r_1, that

(7.37) $$\phi_{q_1 q_1} = \phi_{q_1 r_1} = 0.$$

Also,

(7.38) $$\Omega_{q_1} = f_{p_1} f_z + f_{z_1} - f f_{p_1 z} - p_1 f_{p_1 z_1} - r_1 f_{p_1 p_1} + \phi_{r_1} = 0$$

[5] G. L. Lamb, Jr., *Bäcklund transformations for certain nonlinear evolution equations*, J. Mathematical Phys., 15(1974), p. 2157.

holds, as well as

(7.39) $$\Omega_{q_1q_1} = -f_{p_1p_1} + \phi_{r_1r_1} = 0.$$

From (7.34a) it follows that

(7.40) $$f_{p_1p_1} = \phi_{r_1r_1} = a(z, z_1, p_1),$$

where a is to be determined. If this function were nonzero, a (nonlinear) dependence of f upon p_1 would result. This might well lead to a valid transformation, but it would destroy the expected symmetry of (7.34a) as well as the anticipated relation to the Riccati equation. In what follows we shall consider only that case in which $a \equiv 0$.

With $a \equiv 0$, equation (7.40) yields

(7.41) $$f(z, z_1, p_1) = b(z, z_1)p + c(z, z_1),$$

(7.42) $$\phi(z, z_1, q_1, p_1, r_1) = b(z, z_1)q_1 + \lambda(z, z_1, p_1)r_1 + \nu(z, z_1, p_1),$$

in which (7.37) and (7.36) have been employed to obtain further simplifications. The functions b, c, λ, and ν arise in the integration process and are to be determined. From (7.35) one also finds $\Omega_{r_1r_1} = -2\phi_{r_1p_1} = 0$, so that λ must be independent of p_1. Also it is easily seen that $\Omega_{p_1p_1p_1} = 0$, which yields

(7.43) $$\nu(z, z_1, p_1) = \nu_2(z, z_1)p_1^2 + \nu_1(z, z_1)p_1 + \nu_0(z, z_1).$$

Useful results are obtained by assuming $b(z, z_1)$ to be a constant, denoted by b, although other transformations result if this assumption is not made. Under this assumption the transformation reduces to

(7.44a) $$p = bp_1 + c(z, z_1),$$

(7.44b) $$q = bq_1 + \lambda r_1 + \nu_2 p_1^2 + \nu_1 p_1 + \nu_0,$$

where c, λ, ν_2, ν_1, and ν_0 are undetermined functions of z and z_1. The expected symmetry of the first equation in p_1 and the second in q_1 is realized.

The explicit dependence upon r_1, q_1, and p_1 has now been obtained. Therefore, upon substitution of (7.44) into (7.35), it follows that the coefficients of r_1, q_1, and p_1, etc., must vanish (since $\Omega_{r_1} = \Omega_{p_1} = \Omega_{q_1} = 0$). From this we obtain

(7.45a) $$2\nu_2 = -(b\lambda_z + \lambda_{z_1}),$$

(7.45b) $$\lambda = -(bc_z + c_{z_1}),$$

(7.45c) $$\nu_1 = \lambda c_z - c\lambda_z,$$

(7.45d) $$\nu_2 c_z - c\nu_{2z} + 3\lambda - b\nu_{1z} - \nu_{1z_1} = 0,$$

(7.45e) $$\nu_1 c_z - c\nu_{1z} - \nu_{0z_1} - b\nu_{0z} = 0,$$

(7.45f) $$b\nu_{2z} + \nu_{2z_1} = 0,$$

(7.45g) $$\nu_0 c_z - c\nu_{0z} = 0,$$

which constitute seven equations in the five unknown functions λ, c, ν_0, ν_1, ν_2

and the constant b. Such overdetermined systems are typical of calculations dealing with Bäcklund transformations.

Previously it was required that z_1 satisfy (7.33). It is now also required that z satisfy the same equation. Calculation of $\alpha \equiv z_{xxx}$ from (7.44a), using (7.45a, b), gives

(7.46) $$\alpha = b\alpha_1 - \lambda r_1 + 2\nu_2 p_1^2 + p_1\left[2bcc_{zz} + 2cc_{zz_1} + c_z\left(bc_z + c_{z_1}\right) + c^2c_{zz} + cc_z^2\right].$$

Since z must satisfy $q + 3p^2 + \alpha = 0$, there follows

(7.47a) $$\nu_2 = b - b^2,$$

(7.47b) $$bc_{zz} + 2b + c_{zz_1} = 0,$$

(7.47c) $$c^2c_{zz} + cc_z^2 + \nu_0 + 3c^2 = 0.$$

Equation (7.45g) implies

(7.48) $$\nu_0 = \psi\,(z_1)c\,(z, z_1),$$

where $\psi\,(z_1)$ is to be found. Substituting ν_0 into (7.47c) and integrating once, we find

(7.49) $$c_z^2 + 2c + \psi + Kc^{-2} = 0.$$

To avoid elliptic functions consider only $K = 0$, whereupon

(7.50) $$c_{zz} = -1.$$

Consequently, from (7.47b),

(7.51) $$c_{zz_1} = -b$$

and finally, from (7.47a) plus (7.45a, b),

(7.52) $$c_{z_1z_1} = 2b + b^2.$$

Integration of these last three equations yields

(7.53) $$c\,(z, z_1) = m - \tfrac{1}{2}\left[z^2 + 2bzz_1 - b\,(2+b)z_1^2\right] + kz + lz_1,$$

where k, l, and m are arbitrary constants. Since the derivative of z yields the desired u (see first paragraph of this section), solutions with $m \neq 0$ can be used. In what follows we use $m \neq 0$, $k = l = 0$.

From (7.45) we now find

(7.54a) $$\lambda = 2b\,(z - z_1),$$

(7.54b) $$\nu_1 = -2bm - b\left(z^2 - 2zz_1 + b^2z_1^2\right),$$

(7.54c) $$\nu_2 = b - b^2,$$

(7.54d) $$\nu_0 = \left[-2m - 2b\,(1+b)z_1^2\right]c\,(z, z_1).$$

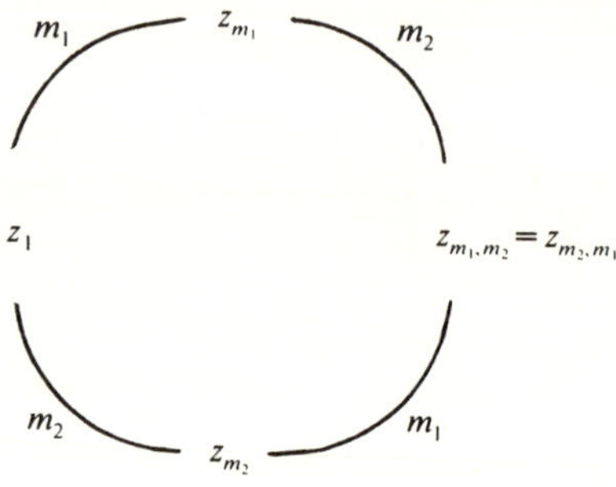

FIG. 7. Diagram for (7.56).

Equations (7.45a, b, c, f, g) are satisfied identically by these, while (7.45d, e) require $b=-1$. Hence an invariance Bäcklund transformation for the transformed KdV equation (7.33) is

(7.55a) $$p+p_1=m-\tfrac{1}{2}(z-z_1)^2,$$

(7.55b) $$q+q_1=(z-z_1)(r-r_1)-2\left(p^2+pp_1+p_1^2\right),$$

where m is an arbitrary constant. This result was also obtained by Wahlquist and Estabrook[6] by an alternative process.

A theorem of permutability (nonlinear superposition) which permits the iterative construction of an infinite sequence of particular solutions also exists here, provided we know three solutions. This was observed by Wahlquist and Estabrook and has been used in applications by others (see Chapter 4). To obtain the composition relation, (7.55a) is interpreted as a transformation from a known solution z_1 of (7.33) to another solution z_m which is obtained by the use of the constant m. The solution obtained from z_{m_1} by using (7.55a) with m_2 is denoted by z_{m_1,m_2}. Corresponding notation is employed for the p's. Writing four such transformations gives

$$p_{m_1}+p_1=m_1-\tfrac{1}{2}(z_{m_1}-z_1)^2,$$

$$p_{m_2}+p_1=m_2-\tfrac{1}{2}(z_{m_2}-z_1)^2,$$

$$p_{m_1,m_2}+p_{m_1}=m_2-\tfrac{1}{2}(z_{m_1,m_2}-z_{m_1})^2,$$

$$p_{m_1,m_2}+p_{m_2}=m_1-\tfrac{1}{2}(z_{m_1,m_2}-z_{m_2})^2,$$

where the two already calculated solutions z_{m_1} and z_{m_2} are used in the last two equations. The last two equations must yield the same final solution (i.e., $z_{m_1,m_2}=z_{m_2,m_1}$, as has already been stated; this is demonstrated in the diagram

[6]H. D. Wahlquist and F. B. Estabrook, *Bäcklund transformation for solutions of the Korteweg-de Vries equation*, Phys. Rev. Lett., 31(1973), p. 1386.

(Fig. 7) of the type used by Bianchi. To obtain a nonlinear superposition subtract the second equation from the first, the fourth from the third and eliminate $p_{m_1}-p_{m_2}$ from the two resulting equations to obtain

$$z_{m_1,m_2}=z_1+\frac{2(m_1-m_2)}{z_{m_1}-z_{m_2}}, \tag{7.56}$$

which is a remarkable relation in its simplicity. It can be used to recursively generate an infinite sequence of particular solutions for (7.33).

The second relation, (7.55b), does not yield as elegant a result.

A relation to the Riccati equation is obtainable in a manner similar to the preceding section by setting $r=z-z_1$ in (7.55a), whereupon that equation becomes

$$r_x-\tfrac{1}{2}r^2+(m-2p)=0.$$

Thus we have demonstrated that the nonlinear superposition is associated once again with a Riccati equation.

Modified KdV Equation. Several authors, including Lamb[7], have found that the modified KdV equation

$$u_y+6u^2u_{xx}+u_{xxx}=0,$$

transformed by means of $z=\int_{-\infty}^{x}u(\eta,y)\,d\eta$ to[8]

$$q+2p^3+\alpha=0, \tag{7.57}$$

has a completely symmetric Bäcklund transformation. Without presenting the detailed computations, one finds that the transformation for (7.57) is

$$p=bp_1+a\sin(z+bz_1) \tag{7.58a}$$

$$q=bq_1-a\left[(r+br_1)\cos(z+bz_1)+(p^2+p_1^2)\sin(z+bz_1)\right], \tag{7.58b}$$

where z and z_1 both satisfy (7.57) and $b=\pm 1$. Equation (7.58a) is quite similar to the Bäcklund transformation equation for the sine–Gordon equation [compare (7.28)]. Since only one of the transformation equations is required for the nonlinear superposition, there is a corresponding similarity with that relation as well. By a calculation analogous to those previously given one finds

$$\tan\tfrac{1}{2}(z_4-z_1)=b\left(\frac{a_2+a_3}{a_2-a_3}\right)\tan\tfrac{1}{2}(z_2-z_3) \tag{7.59}$$

with $b=\pm 1$ and $a_2\neq a_3$. A diagram like those used by Bianchi can be drawn.

The change of variable

$$R=\tan\tfrac{1}{2}(z+z_1)$$

[7] Lamb, op. cit.

[8] The term p^n, $n>3$, does not lead to a Bäcklund transformation of "Riccati type."

converts (7.58a) into the Riccati equation

$$R_x + aR - p(1+R^2) = 0.$$

A Nonlinear Schrödinger Equation. The equation under study and its complex conjugate are taken in the form

$$iq + r + z^2\bar{z} = 0, \tag{7.60a}$$

$$-i\bar{q} + \bar{r} + \bar{z}^2 z = 0, \tag{7.60b}$$

where the bars indicate complex conjugates. Lamb[9] has derived an invariance Bäcklund transformation for (7.60) beginning with

$$p = f(z, \bar{z}, z_1, \bar{z}_1, p_1, \bar{p}_1),$$
$$q = \phi(z, \bar{z}, z_1, \bar{z}_1, q_1, \bar{q}_1, p_1, \bar{p}_1).$$

The resulting transformation is

$$\begin{aligned} p &= p_1 - \tfrac{1}{2} iw\tau + ikv, \\ q &= q_1 + \tfrac{1}{2}\tau(p + p_1) - kn + \tfrac{1}{4} iv(|w|^2 + |v|^2), \end{aligned} \tag{7.61}$$

where

$$\tau = \pm i(b - 2|v|^2)^{1/2},$$
$$w = z + z_1,$$
$$v = z - z_1,$$
$$n = -\tfrac{1}{2} iw\tau + ikv,$$

and b and k are arbitrary real constants.

Setting

$$R = (b - 2|v|^2)^{1/2} / v\sqrt{2}$$

both (7.61) and its complex conjugate become

$$z\left[R_x + ikR + 2^{-1/2}(zR^2 + \bar{z})\right] = z_1\left[R_x + ikR + 2^{-1/2}(z_1 R^2 + \bar{z}_1)\right].$$

If one now sets either z or z_1 equal to zero, the other variable satisfies a Riccati equation. Thus there is a nonlinear superposition, but it is very complicated.

§8. Transformations Relating Different Differential Equations

The Bäcklund transformations of §7 are of a special type, which we have called invariance transformations because they transform a given equation into itself. A more general use of the theory involves transformations between equations of *different* form. Some of these relate the solutions of a nonlinear

[9] Lamb, op. cit.

equation to those of a linear equation. Transformations of the Liouville equation $u_{xy}=e^u$ and the Burgers equation $u_y+uu_x=u_{xx}$ are of this nonlinear–linear type. Others relate two nonlinear equations wherein the solutions of the second are more easily obtained. Both processes are very useful, but unfortunately they had been used only sparingly until a very recent (1977) upsurge in interest. In what follows we will present detailed examples of both kinds.

Liouville's Equation. Liouville's equation

$$(z_1)_{xy}=e^{z_1} \tag{8.1}$$

provides an excellent simple example of an equation for which a Bäcklund transformation exists relating z_1 satisfying (8.1) to z satisfying the linear equation

$$z_{xy}=0. \tag{8.2}$$

As in the sine–Gordon equation, consider the special case

$$p=f(p_1,z,z_1), \qquad q=\psi(q_1,z,z_1),$$

where z_1 satisfies (8.1). The integrability requirement $dp/dy=dq/dx$ generates the relation

$$\Omega=e^{z_1}\left(\frac{\partial f}{\partial p_1}-\frac{\partial \psi}{\partial q_1}\right)-p_1\frac{\partial \psi}{\partial z_1}+q_1\frac{\partial f}{\partial z_1}+\psi\frac{\partial f}{\partial z}-f\frac{\partial \psi}{\partial z}=0. \tag{8.3}$$

Calculation of the derivatives of Ω to that point where z_1 is no longer explicitly present yields the equations

$$\frac{\partial \Omega}{\partial p_1}=e^{z_1}\frac{\partial^2 f}{\partial p_1^2}-\frac{\partial \psi}{\partial z_1}+q_1\frac{\partial^2 f}{\partial p_1\,\partial z_1}+\psi\frac{\partial^2 f}{\partial p_1\,\partial z}-\frac{\partial f}{\partial p_1}\frac{\partial \psi}{\partial z}=0, \tag{8.4}$$

$$\frac{\partial \Omega}{\partial q_1}=-e^{z_1}\frac{\partial^2 \psi}{\partial q_1^2}-p_1\frac{\partial^2 \psi}{\partial q_1\,\partial z_1}+\frac{\partial f}{\partial z_1}+\frac{\partial \psi}{\partial q_1}\frac{\partial f}{\partial z}-f\frac{\partial^2 \psi}{\partial q_1\,\partial z}=0, \tag{8.5}$$

$$\frac{\partial^2 \Omega}{\partial p_1\,\partial q_1}=-\frac{\partial^2 \psi}{\partial q_1\,\partial z_1}+\frac{\partial^2 f}{\partial p_1\,\partial z_1}+\frac{\partial \psi}{\partial q_1}\frac{\partial^2 f}{\partial p_1\,\partial z}-\frac{\partial f}{\partial p_1}\frac{\partial^2 \psi}{\partial q_1\,\partial z}=0, \tag{8.6}$$

$$\frac{\partial^3 \Omega}{\partial p_1^2\,\partial q_1}=\frac{\partial^3 f}{\partial p_1^2\,\partial z_1}+\frac{\partial \psi}{\partial q_1}\frac{\partial^3 f}{\partial p_1^2\,\partial z}-\frac{\partial^2 f}{\partial p_1^2}\frac{\partial^2 \psi}{\partial q_1\,\partial z}=0, \tag{8.7}$$

$$\frac{\partial^3 \Omega}{\partial p_1\,\partial q_1^2}=-\frac{\partial^3 \psi}{\partial q_1^2\,\partial z_1}+\frac{\partial^2 \psi}{\partial q_1^2}\frac{\partial^2 f}{\partial p_1\,\partial z}-\frac{\partial f}{\partial p_1}\frac{\partial^3 \psi}{\partial q_1^2\,\partial z}=0, \tag{8.8}$$

$$\frac{\partial^4 \Omega}{\partial p_1^2\,\partial q_1^2}=\frac{\partial^2 \psi}{\partial q_1^2}\frac{\partial^3 f}{\partial p_1^2\,\partial z}-\frac{\partial^2 f}{\partial p_1^2}\frac{\partial^3 \psi}{\partial q_1^2\,\partial z}=0. \tag{8.9}$$

Equation (8.9) satisfies the requirement that it be free of explicit dependence upon z_1, although solutions of that equation will depend upon z_1 parametrically.

Since these are the same as those for the sine–Gordon equation [(7.4)–(7.10)], *except that* $\sin z_1$ *is replaced by* e^{z_1}, the same analysis will apply.

Let us pick up the analysis *after* (7.24) with the same terminology and the understanding that $\sin z_1$ is replaced by e^{z_1} in all the pertinent equations. Thus (8.4) yields

$$(\bar{r}-r)e^{z_1}+\mu\frac{\partial m}{\partial z}-m\frac{\partial \mu}{\partial z}=0. \tag{8.10}$$

With $\bar{r}=1$ and $r=-1$, equations (7.22) and (7.24) have the simple solutions

$$m=m\,(z+z_1),\qquad \mu=\mu\,(z-z_1). \tag{8.11}$$

With these, (8.10) becomes

$$2e^{z_1}+\frac{\partial}{\partial z_1}\left[m\,(z+z_1)\mu\,(z-z_1)\right]=0. \tag{8.12}$$

Upon differentiation with respect to z there results

$$\frac{m''\,(z+z_1)}{m\,(z+z_1)}=\frac{\mu''\,(z-z_1)}{\mu\,(z-z_1)}=K^2, \tag{8.13}$$

where real exponential solutions are anticipated by the choice of sign in (8.13), and K is a (positive) real constant. The classical solutions to (8.13) are

$$m=A\exp\left[K\,(z+z_1)\right]+B\exp\left[-K\,(z+z_1)\right],$$

$$\mu=\alpha\exp\left[K\,(z-z_1)\right]+\beta\exp\left[-K\,(z-z_1)\right].$$

Upon substituting these in (8.12) and requiring that equation to be identically satisfied, we find

$$K=\tfrac{1}{2},\qquad A\beta=-2,\quad\text{and}\quad \alpha B=0. \tag{8.14}$$

With the simple, but not unique, choices

$$A=a\ (\text{const})\neq 0,\qquad \beta=-2/a,$$

$$\alpha=0,\qquad aB=0,$$

the Bäcklund transformation becomes

$$p=p_1+a\exp\left[\tfrac{1}{2}(z+z_1)\right], \tag{8.15a}$$

$$q=-q_1-\frac{2}{a}\exp\left[-\tfrac{1}{2}(z-z_1)\right]. \tag{8.15b}$$

Two elementary calculations verify that z_1 satisfies $(z_1)_{xy}=e^{z_1}$ and z satisfies $z_{xy}=0$. When the general solution of $z_{xy}=0$, namely $z=F(x)+G(y)$, F and G arbitrary, is inserted into (8.15), the solutions of the resulting simple first-order equations readily yield the general solution of the Liouville equation. A nonlinear superposition results from either (8.15a) or (8.15b). Using (8.15a), we generate z_2 from z_1 via a_2, z_3 from z_1 via a_3, and then z_4 from z_2 via a_3 and from

z_3 via a_2 to obtain

$$p_2-p_1=a_2\exp\left[\tfrac{1}{2}(z_2+z_1)\right],$$
$$p_3-p_1=a_3\exp\left[\tfrac{1}{2}(z_3+z_1)\right],$$
$$p_4-p_2=a_3\exp\left[\tfrac{1}{2}(z_4+z_2)\right],$$
$$p_4-p_3=a_2\exp\left[\tfrac{1}{2}(z_4+z_3)\right].$$

When p_2-p_3 is obtained from the first pair and equated to the same result from the second pair, the nonlinear superposition

(8.16)
$$\frac{a_2}{a_3}\left\{e^{\frac{1}{2}(z_2+z_1)}-e^{\frac{1}{2}(z_4+z_3)}\right\}=e^{\frac{1}{2}(z_3+z_1)}-e^{\frac{1}{2}(z_4+z_2)}$$

results.

Of course, an invariance transformation could also have been developed by a slight modification of the argument following (8.13).

KdV and Modified KdV Equations. The relation between the KdV equation

(8.17)
$$q_1+6z_1p_1+\alpha_1=0$$

and the modified KdV equation

(8.18)
$$q-6z^2p+\alpha=0,$$

discovered by Miura[10], led Lamb[11] to suspect that the two were related by a Bäcklund transformation. Equations (8.17) and (8.18) are taken in the present form to avoid the introduction of complex quantities. Motivated by the Miura transformation $\pm p=z_1+z^2$, the trial form of the transformation will be

(8.19a)
$$p=f(z,z_1),$$

(8.19b)
$$q=\eta(z,z_1,p_1,r_1).$$

The choice of the second equation is motivated by the type of derivatives that appear when equality of mixed second partial derivatives is imposed together with the structure of the present equations, (8.17) and (8.18).

From the integrability condition there follows

(8.20)
$$\Omega=f_z\psi+f_{z_1}q_1-f\psi_z-p_1\psi_{z_1}-r_1\psi_{p_1}+\psi_{r_1}(q_1+6z_1p_1)=0.$$

[10]R. M. Miura, *Korteweg-de Vries equation and generalizations. I. A remarkable explicit nonlinear transformation*, J. Mathematical Phys., 9(1968), p. 1202.

[11]Lamb, op. cit.

Then by direct calculation

$$(8.21) \qquad \Omega_{q_1} = f_{z_1} + \psi_{r_1} = 0,$$

$$(8.22) \qquad \Omega_{p_1} = f_z \psi_{p_1} - f\psi_{zp_1} - \psi_{z_1} - p_1 \psi_{z_1 p_1} - r_1 \psi_{p_1 p_1} + \psi_{r_1 p_1}(q_1 + 6z_1 p_1) + 6z_1 \psi_{r_1} = 0,$$

$$(8.23) \qquad \Omega_{q_1 p_1} = \psi_{r_1 p_1} = 0,$$

$$(8.24) \qquad \Omega_{p_1 r_1} = f_{z_1 z_1} - \psi_{p_1 p_1} = 0,$$

$$(8.25) \qquad \Omega_{p_1 p_1} = f_z f_{z_1 z_1} - f f_{z z_1 z_1} - 2\psi_{z_1 p_1} - p_1 f_{z_1 z_1 z_1} = 0,$$

$$(8.26) \qquad \Omega_{p_1 r_1 p_1} = -\psi_{p_1 p_1 p_1} = 0,$$

$$(8.27) \qquad \Omega_{p_1 p_1 p_1} = -3 f_{z_1 z_1 z_1} = 0.$$

From (8.27) and (8.24) it follows that

$$(8.28) \qquad f_{z_1 z_1} = \psi_{p_1 p_1} = a(z),$$

where $a(z)$ is an unknown function arising from the integration of (8.27). To have (8.19a) agree with the Miura transformation, $a(z)$ must be identically zero. With that, (8.28) and (8.21) yield

$$(8.29) \qquad f_{z_1} = g(z) = -\psi_{r_1}$$

and

$$\psi_{p_1} = \lambda(z, z_1, r_1),$$

where $g(z)$ and $\lambda(z, z_1, r_1)$ are to be determined. Equation (8.23) implies λ does not depend on r_1, and (8.25) forces λ to be independent of z_1. Hence

$$\psi_{p_1} = \lambda(z).$$

Integration of this result and the application of (8.21) and (8.29) leads to

$$(8.30) \qquad \psi = \lambda(z) p_1 - g(z) r_1 + (z, z_1)$$

as our candidate. The z_1 dependence is obtained from the observations that

$$(8.31) \qquad \begin{aligned} \Omega_{z_1 z_1 p_1} &= -\psi_{z_1 z_1 z_1} = 0, \\ \Omega_{z_1 z_1 z_1} &= 3\left[f_{z z_1} \psi_{z_1 z_1} - \psi_{z z_1 z_1} f_{z_1} \right] = 0. \end{aligned}$$

These results are obtained using the previously derived properties—$\psi_{z_1 p_1} = 0$ etc. —where necessary. Since $f_{z_1} = g(z)$, integration of (8.31) yields

$$(8.32) \qquad \psi_{z_1 z_1} = Kg(z),$$

where K is an integration constant. Two integrations of (8.32) yield

$$\psi = \tfrac{1}{2} Kg(z)(z_1)^2 + k(z) z_1 + l(z) + \lambda p_1 - g(z) r_1,$$

which is a form of ψ in which all but the z dependence has been determined.

The Bäcklund transformation is now in the form

(8.33a) $$p=g(z)z_1+h(z),$$

(8.33b) $$q=\lambda(z)p_1-g(z)r_1+\tfrac{1}{2}Kg(z)(z_1)^2+k(z)z_1+l(z),$$

where g, h, λ, k, and l are undetermined functions of z. Substituting (8.33) in (8.20), and then setting equal to zero the coefficient of each of the different terms of Ω which depend upon z_1 and its derivatives, we have

(8.34a) $$g\lambda'-\lambda g'+g(6+K)=0,$$

(8.34b) $$g'k-k'g+\tfrac{1}{2}K(gh'-hg')=0,$$

(8.34c) $$g'l-l'g+kh'-hk'=0,$$

(8.34d) $$k=\lambda h'-h\lambda',$$

(8.34e) $$lh'-hl'=0,$$

(8.34f) $$\lambda=hg'-gh'.$$

The combination of (8.34b) and (8.34f) immediately yields

(8.35) $$K\lambda=2(kg'-gk').$$

Specialization of (8.33a) to the Miura transformation requires $g=\pm1$. The construction of Ω has already used the KdV equation. For z to satisfy the modified KdV equation $q-6z^2p+\alpha=0$, we must have from (8.33)

(8.36a) $$h'+\lambda=0,$$

(8.36b) $$h''+\tfrac{1}{2}K=0,$$

(8.36c) $$2hh''+(h')^2+k-6z^2=0,$$

(8.36d) $$h''h^2+(h')^2h+l-6z^2h=0.$$

The two sets of equations, (8.34) and (8.36), when solved yield the following results:

$$g=\pm1,\quad K=-4,\quad h=\pm z^2,\quad \lambda=-2z,\quad k=\mp 2z^2,\quad l=0.$$

Thus *one* Bäcklund transformation relating the KdV equation $q_1+6z_1p_1+\alpha_1=0$ to the modified equation $q-6z^2p+\alpha=0$ is

(8.37a) $$p=\pm(z_1+z^2),$$

(8.37b) $$q=\mp r_1-2(zp_1+z_1p).$$

Clearly (8.37a) is the Miura transformation, while (8.37b) furnishes the companion equation which completes the Bäcklund transformation.

The assumptions and restrictions of the preceding analysis clearly show the nonuniqueness of these transformations. Further exploration may uncover transformations relating the original equation to much simpler forms.

Burgers Equation. The Burgers equation

$$q_1 + z_1 p_1 + r_1 = 0 \tag{8.38}$$

arises in turbulence and wave propagation. In 1906[12] it was already known to possess a relationship with the linear equation $q+r=0$. (This was rediscovered by Hopf and Cole some fifty years later. For a discussion of those results and a simple explanation of them see Ames [1], pp. 8–10 and 41–49.)[13]

A Bäcklund transformation relating (8.38) and the linear diffusion equation

$$q + r = 0 \tag{8.39}$$

is

$$p = \tfrac{1}{2} z z_1, \tag{8.40a}$$

$$q = -\tfrac{1}{2}(z p_1 + p z_1). \tag{8.40b}$$

Equation (8.40a) is the Forsyth relation, while (8.40b) completes the Bäcklund-transformation pair.

Example of Clairin. The equations $s_1 = 0$ and $s + q e^{-z} = 0$ are transformed into one another by the Bäcklund transformation

$$p = p_1 + e^{-z},$$
$$q = e^{z_1 - z}.$$

[12]See Forsyth, op. cit.

[13]See also Volume I, p. 23, of this set.

Chapter **2**

Tangent Transformation Groups

As discussed in Chapter 1 and demonstrated in the classical literature, there exist no invertible higher-order tangent transformations whose action is closed in any finite-dimensional space other than prolongations of Lie point or tangent transformations (see §§2, 3), yet there exist ∞-valued, higher-order tangent surface transformations (Bäcklund transformations) in such finite-dimensional spaces (see §§5, 6). The existence of these transformations finds many fruitful applications in geometry and the analysis of differential equations (see §§5–8 and Chapter 4), but their noninvertible form does not lead to a generalization of Lie's theory of the group analysis of differential equations founded on higher-order tangent transformations. This was the state of this transformation theory founded on higher-order tangency until recently.

It is clear from the preceding that if there exists any possibility of realizing invertible higher-order tangent surface-element transformations it will have to arise from the action of point transformations on the ∞-order surface elements $(x, u, \underset{1}{u}, \ldots)$. The objective of this chapter is to show that it is indeed possible in this way to construct a generalization of Lie's theory of tangent transformation groups and realize its application to the group analysis of differential equations (Ibragimov and Anderson [1]). Further, this generalization is necessarily founded upon ∞-order tangent transformation groups (Lie-Bäcklund transformation groups) acting in infinite-dimensional spaces. Here we confine our attention to continuous transformation groups, and in this way we are able to give an infinitesimal characterization of these groups.

In the interests of presenting a self-contained discussion and in order to establish terminology and basic techniques, we begin with a discussion of the tangent transformation groups of Sophus Lie. A proof of Bäcklund's result on the nonexistence of finite-order tangent transformations beyond prolongations of Lie's tangent transformations (see §2) is reproduced here for tangent transformation groups using infinitesimal techniques. This clearly delineates the scope of applicability of Lie tangent transformation groups and sets the stage for the introduction of Lie-Bäcklund transformation groups. An infinitesimal characterization of these transformation groups is given, as well as a proof that when they represent a generalization of Lie transformation groups they are point transformation groups acting in an infinite-dimensional space. Finally, the problems of establishing the integrability of the corresponding Lie-Bäcklund equations and the equivalence structure of these transformation groups are treated.

I. FINITE-ORDER TANGENT TRANSFORMATIONS

§9. Tangent Transformation Groups of Sophus Lie

Consider the group G of point transformations

$$G:\begin{cases} x'^i=f^i(x,u,\underset{1}{u};a), & i=1,\cdots,n,\\ u'^\alpha=\phi^\alpha(x,u,\underset{1}{u};a), & \alpha=1,\cdots,m,\\ u_i'^\alpha=\psi_i^\alpha(x,u,\underset{1}{u};a), & i=1,\cdots,n,\quad \alpha=1,\cdots,m,\end{cases} \tag{9.1}$$

in the space of independent variables $(x, u, \underset{1}{u})$—where a is the group parameter, $x=(x^1,\cdots,x^n)\in\mathbb{R}^n$, $u=(u^1,\cdots,u^m)\in\mathbb{R}^m$, and $\underset{1}{u}=(u_1^1,\ u_2^1,\cdots,u_n^m)\in\mathbb{R}^{nm}$—together with another group $\tilde{G}$ of point transformations in the space of independent variables $(x, u, \underset{1}{u}, dx, du, d\underset{1}{u})$. The group $\tilde{G}$ is obtained by the extension of the action of the group G to the differentials by means of the formulas

$$\begin{aligned} dx'^i&=\frac{\partial f^i}{\partial x^j}dx^j+\frac{\partial f^i}{\partial u^\beta}du^\beta+\frac{\partial f^i}{\partial u_j^\beta}du_j^\beta,\\ du'^\alpha&=\frac{\partial \phi^\alpha}{\partial x^j}dx^j+\frac{\partial \phi^\alpha}{\partial u^\beta}du^\beta+\frac{\partial \phi^\alpha}{\partial u_j^\beta}du_j^\beta,\\ du_i'^\alpha&=\frac{\partial \psi_i^\alpha}{\partial x^j}dx^j+\frac{\partial \psi_i^\alpha}{\partial u^\beta}du^\beta+\frac{\partial \psi_i^\alpha}{\partial u_j^\beta}du_j^\beta.\end{aligned} \tag{9.2}$$

S. Lie treated the case $m=1$ and called the group G a group of tangent transformations if the equation

$$du-u_i\,dx^i=0, \tag{9.3}$$

is invariant with respect to the extended group $\tilde{G}$. Hereafter, we only refer to a transformation group of this type as *a group of Lie tangent transformations*.

A natural direction in which to attempt to generalize the notion of the tangent transformations of S. Lie is to consider the case of arbitrary m. Indeed, we say that G is a group of tangent transformations if the equations

$$du^\alpha - u_i^\alpha dx^i = 0, \qquad \alpha = 1, \cdots, m, \tag{9.4}$$

are invariant with respect to the group $\tilde{G}$. One example of such a group of tangent transformations is the group $\mathbb{P}$ of Lie point transformations in the (x, u)-space

$$\mathbb{P}: \begin{cases} x'^i = g^i(x, u; a), & i = 1, \cdots, n, \\ u'^\alpha = h^\alpha(x, u; a), & \alpha = 1, \cdots, m, \end{cases} \tag{9.5}$$

extended in the usual way (Lie [1], Eisenhart [1], Ovsjannikov [2],[4]) to "derivatives" $\underset{1}{u}$. The following theorem shows that such an extended group of Lie point transformations is the only possible type of tangent transformation group for $m > 1$. This result was first brought to the attention of one of the authors (N.H.I.) by P. Kucharczyk in 1965, and the proof was communicated in the same year to the same author by L. V. Ovsjannikov (see Ibragimov [1]).

THEOREM 9.1. *If* $m > 1$, *then every group* G *of tangent transformations* (9.1) *is the extension of a group* $\mathbb{P}$ *of Lie point transformations* (9.5) *to the derivatives* $\underset{1}{u}$.

Proof. To prove this statement the infinitesimal criterion for the invariance of (9.4) is used. Let

$$X = \xi^i \frac{\partial}{\partial x^i} + \eta^\alpha \frac{\partial}{\partial u^\alpha} + \zeta_i^\alpha \frac{\partial}{\partial u_i^\alpha} \tag{9.6}$$

be the infinitesimal operator of the group G where

$$\xi^i = \left.\frac{\partial f^i}{\partial a}\right|_{a=0}, \quad \eta^\alpha = \left.\frac{\partial \phi^\alpha}{\partial a}\right|_{a=0}, \quad \zeta_i^\alpha = \left.\frac{\partial \psi_i^\alpha}{\partial a}\right|_{a=0},$$

and $i = 1, \cdots, n$, $\alpha = 1, \cdots, m$. The infinitesimal operator $\tilde{X}$ of the group $\tilde{G}$ will have the form (the operator of differentiation with respect to du_i^α is not needed here and is omitted)

$$\tilde{X} = X + \tilde{\xi}^i \frac{\partial}{\partial (dx^i)} + \tilde{\eta}^\alpha \frac{\partial}{\partial (du^\alpha)}, \tag{9.7}$$

where

$$\tilde{\xi}^i = \left.\frac{\partial (dx'^i)}{\partial a}\right|_{a=0}, \qquad \tilde{\eta}^\alpha = \left.\frac{\partial (du'^\alpha)}{\partial a}\right|_{a=0}.$$

Using (9.2) and the definition of the quantities ξ^i, η^α, the coordinates $\tilde{\xi}^i$, $\tilde{\eta}^\alpha$ can be written in the form

$$(9.8)\qquad \begin{aligned}\tilde{\xi}^i &= \frac{\partial \xi^i}{\partial x^j} dx^j + \frac{\partial \xi^i}{\partial u^\alpha} du^\alpha + \frac{\partial \xi^i}{\partial u_i^\alpha} du_i^\alpha, \qquad i=1,\cdots,n,\\ \tilde{\eta}^\alpha &= \frac{\partial \eta^\alpha}{\partial x^j} dx^j + \frac{\partial \eta^\alpha}{\partial u^\beta} du^\beta + \frac{\partial \eta^\alpha}{\partial u_i^\beta} du_i^\beta, \qquad \alpha=1,\cdots,m.\end{aligned}$$

The infinitesimal criterion for the invariance (Ovsjannikov [2]) of equations (9.4) is

$$\tilde{X}\left(du^\alpha - u_i^\alpha dx^i\right)\big|_{(9.4)}=0, \qquad \alpha=1,\cdots,m,$$

or, from (9.6)–(9.8) and after a suitable rearrangement,

$$\left(\frac{\partial \eta^\alpha}{\partial x^j} + \frac{\partial \eta^\alpha}{\partial u^\beta} u_j^\beta - u_i^\alpha \frac{\partial \xi^i}{\partial x^j} - u_i^\alpha u_j^\beta \frac{\partial \xi^i}{\partial u^\beta} - \zeta_j^\alpha\right) dx^j + \left(\frac{\partial \eta^\alpha}{\partial u_j^\beta} - u_i^\alpha \frac{\partial \xi^i}{\partial u_j^\beta}\right) du_j^\beta = 0,$$
$$\alpha=1,\cdots,m.$$

In this equation all the quantities dx^j, du_j^β are independent, which implies that the coefficients of these terms are each separately zero. This yields the following system of partial differential equations for the functions $\xi^i(x, u, \underset{1}{u})$, $\eta^\alpha(x, u, \underset{1}{u})$, $\zeta_i^\alpha(x, u, \underset{1}{u})$:

$$(9.9)\qquad \zeta_i^\alpha = \frac{\partial \eta^\alpha}{\partial x^i} + \frac{\partial \eta^\alpha}{\partial u^\beta} u_i^\beta - u_j^\alpha \left(\frac{\partial \xi^j}{\partial x^i} + u_i^\beta \frac{\partial \xi^j}{\partial u^\beta}\right), \qquad \frac{\partial \eta^\alpha}{\partial u_i^\beta} - u_j^\alpha \frac{\partial \xi^j}{\partial u_i^\beta} = 0,$$

where $i=1,\cdots,n$, α, $\beta=1,\cdots,m$. After introducing the functions

$$(9.10)\qquad W^\alpha = \eta^\alpha - \xi^i u_i^\alpha, \qquad \alpha=1,\cdots,m,$$

these equations can be rewritten as

$$(9.11)\qquad \zeta_i^\alpha = \frac{\partial W^\alpha}{\partial x^i} + u_i^\beta \frac{\partial W^\alpha}{\partial u^\beta}, \qquad \xi^i \delta_\beta^\alpha + \frac{\partial W^\alpha}{\partial u_i^\beta} = 0.$$

The second relation in (9.11) implies for $m>1$ that

$$\frac{\partial W^\alpha}{\partial u_i^\beta} = 0, \qquad \alpha \neq \beta,$$

$$\frac{\partial W^1}{\partial u_i^1} = \frac{\partial W^2}{\partial u_i^2} = \cdots = \frac{\partial W^m}{\partial u_i^m}.$$

These equations have the general solution

$$W^\alpha = U^\alpha(x, u) - V^i(x, u) u_i^\alpha.$$

After the substitution of this general solution into (9.11) and (9.10), we obtain

$$\xi^i = V^i\,(x, u),$$
$$\eta^\alpha = U^\alpha\,(x, u),$$
$$\zeta_i^\alpha = D_i\,(\eta^\alpha) - u_j^\alpha D_i\,(\xi^j),$$

where

$$D_i = \frac{\partial}{\partial x^i} + u_i^\alpha \frac{\partial}{\partial u^\alpha}, \qquad i = 1, \cdots, n.$$

These are precisely the formulas for the coordinates of the extended infinitesimal operator of the group $\mathbb{P}$ of Lie point transformations (9.5). ■

Equation (9.11) in the preceding proof is valid for arbitrary m, in particular for $m=1$. This, together with (9.10), immediately yields Lie's infinitesimal characterization of his tangent transformations, which can be expressed as the following proposition (Lie [5] vol. 4, Eisenhart [1]).

THEOREM 9.2. *If $m=1$, the transformations* (9.1) *form a group G of Lie tangent transformations if there exists a function $W\,(x, u, \underset{1}{u})$ such that*

$$\xi^i = -\frac{\partial W}{\partial u_i}, \quad \eta = W - u_i \frac{\partial W}{\partial u_i}, \quad \zeta_i = \frac{\partial W}{\partial x^i} + u_i \frac{\partial W}{\partial u}. \tag{9.12}$$

Summarizing the results thus far, we have the situation that nontrivial (i.e., more general than extensions of Lie point transformations $\mathbb{P}$) tangent transformations exist only for the case of one function u ($m=1$).

§10. Higher-Order Tangent Transformation Groups

Another suggested direction of generalization of Lie's definition of a first-order tangent transformation is to introduce higher-order tangent transformations. Therefore let us consider the group $\underset{k}{G}$ of transformations

$$\underset{k}{G}: \quad \begin{aligned} x'^i &= f^i\left(x, u, \underset{1}{u}, \cdots, \underset{k}{u}; a\right), \\ u'^\alpha &= \phi^\alpha\left(x, u, \underset{1}{u}, \cdots, \underset{k}{u}; a\right), \\ u_i'^\alpha &= \psi_i^\alpha\left(x, u, \underset{1}{u}, \cdots, \underset{k}{u}; a\right), \\ &\vdots \\ u'^\alpha_{i_1 \cdots i_k} &= \psi^\alpha_{i_1 \cdots i_k}\left(x, u, \underset{1}{u}, \cdots, \underset{k}{u}; a\right) \end{aligned} \tag{10.1}$$

$$(i, i_1, \cdots, i_k = 1, \cdots, n; \quad \alpha = 1, \cdots, m)$$

in the space of variables $(x, u, \underset{1}{u}, \cdots, \underset{k}{u})$, where

$$\underset{s}{u} = \{u^{\alpha}_{i_1\cdots i_s} | \alpha = 1, \cdots, m;\ i_1, \cdots, i_s = 1, \cdots, n\} \qquad \text{for } s = 1, \cdots, k.$$

(The quantities $u_{i_1\cdots i_s}$ are taken here to be symmetric in their lower indices.) In this case, we say that $\underset{k}{G}$ is a group of tangent transformations of kth order if the system of equations

$$\begin{aligned} du^{\alpha} - u^{\alpha}_i\, dx^i &= 0, \quad \alpha = 1, \cdots, m, \\ du^{\alpha}_{i_1\cdots i_s} - u^{\alpha}_{i_1\cdots i_s j}\, dx^j &= 0, \quad \alpha = 1, \cdots, m, \quad s = 1, \cdots, k-1, \end{aligned} \tag{10.2}$$

is invariant with respect to the group $\underset{k}{\tilde{G}}$ obtained by the extension of the group $\underset{k}{G}$ to the differentials dx^i, $du^{\alpha}, \cdots, du^{\alpha}_{i_1\cdots i_k}$. But, as discussed in §2, it is known that there do not exist nontrivial higher-order tangent transformations (Bäcklund [1], [2]). This result is fundamental to understanding the need for the definition of a group of Lie-Bäcklund tangent transformations introduced in §11. Since the proof of Theorem 9.1 is not directly applicable to Bäcklund's result because the du^{α}_i are no longer independent variables, we shall present a proof of this result in the language of the infinitesimal criterion for the invariance of (10.2). This proof (Ibragimov and Anderson [1]), which is also integral to the results presented in §11, is based on calculational methods developed by L. V. Ovsjannikov and is an extension to the general case of a previous calculation (Ibragimov [2]).

The operator X of the group $\underset{k}{G}$ of tangent transformations (10.1) is given in general by

$$X = \xi^i \frac{\partial}{\partial x^i} + \eta^{\alpha} \frac{\partial}{\partial u^{\alpha}} + \zeta^{\alpha}_i \frac{\partial}{\partial u^{\alpha}_i} + \cdots + \zeta^{\alpha}_{i_1\cdots i_k} \frac{\partial}{\partial u^{\alpha}_{i_1\cdots i_k}}, \tag{10.3}$$

where

$$\xi^i = \left.\frac{\partial f^i}{\partial a}\right|_{a=0}, \quad \eta^{\alpha} = \left.\frac{\partial \phi^{\alpha}}{\partial a}\right|_{a=0}, \cdots, \quad \zeta^{\alpha}_{i_1\cdots i_s} = \left.\frac{\partial \psi^{\alpha}_{i_1\cdots i_s}}{\partial a}\right|_{a=0}, \tag{10.4}$$

$s = 1, \cdots, k$. As in the preceding, the infinitesimal operator of the group $\underset{k}{G}$ is given by

$$\begin{aligned} \tilde{X} = X + \tilde{\xi}^i \frac{\partial}{\partial (dx^i)} + \tilde{\eta}^{\alpha} \frac{\partial}{\partial (du^{\alpha})} + \tilde{\zeta}^{\alpha}_i \frac{\partial}{\partial (du^{\alpha}_i)} + \cdots \\ + \tilde{\zeta}^{\alpha}_{i_1\cdots i_{k-1}} \frac{\partial}{\partial (du^{\alpha}_{i_1\cdots i_{k-1}})} \end{aligned} \tag{10.5}$$

(as before, the operator of differentiation with respect to $du^{\alpha}_{i_1\cdots i_k}$ is not needed for the following considerations and is omitted). From (10.1) and (10.4) it follows that

$$\tilde{\xi}^i = \frac{\partial \xi^i}{\partial x^j} dx^j + \frac{\partial \xi^i}{\partial u^\alpha} du^\alpha + \frac{\partial \xi^i}{\partial u_j^\alpha} du_j^\alpha + \cdots + \frac{\partial \xi^i}{\partial u_{i_1\cdots i_k}^\alpha} du_{i_1\cdots i_k}^\alpha,$$

$$\tilde{\eta}^\alpha = \frac{\partial \eta^\alpha}{\partial x^j} dx^j + \frac{\partial \eta^\alpha}{\partial u^\beta} du^\beta + \frac{\partial \eta^\alpha}{\partial u_j^\beta} du_j^\beta + \cdots + \frac{\partial \eta^\alpha}{\partial u_{i_1\cdots i_k}} du_{i_1\cdots i_k}^\beta, \tag{10.6}$$

$$\tilde{\zeta}_{i_1\cdots i_s}^\alpha = \frac{\partial \zeta_{i_1\cdots i_s}^\alpha}{\partial x^j} dx^j + \frac{\partial \zeta_{i_1\cdots i_s}^\alpha}{\partial u^\beta} du^\beta + \frac{\partial \zeta_{i_1\cdots i_s}^\alpha}{\partial u_j^\beta} du_j^\beta$$

$$+ \cdots + \frac{\partial \zeta_{i_1\cdots i_s}^\alpha}{\partial u_{j_1\cdots j_k}^\beta} du_{j_1\cdots j_k}^\beta,$$

where $s=1,\cdots,k$. Now the criteria for the invariance of equations (10.2) are given by

$$\begin{aligned} &\tilde{X}\left(du^\alpha - u_i^\alpha dx^i\right)\big|_{(10.2)} = 0, \\ &\tilde{X}\left(du_{i_1\cdots i_s}^\alpha - u_{i_1\cdots i_s j}^\alpha dx^j\right)\big|_{(10.2)} = 0, \quad s=1,\cdots,k-1, \end{aligned} \tag{10.7}$$

or from (10.6),

$$\begin{aligned} &\left(\tilde{\eta}^\alpha - \zeta_i^\alpha dx^i - u_i^\alpha \tilde{\xi}^i\right)\big|_{(10.2)} = 0, \\ &\left(\tilde{\zeta}_{i_1\cdots i_s}^\alpha - \zeta_{i_1\cdots i_s j}^\alpha dx^j - u_{i_1\cdots i_s j}^\alpha \tilde{\xi}^j\right)\big|_{(10.2)} = 0, \quad s=1,\cdots,k-1. \end{aligned} \tag{10.8}$$

It is now convenient to express our formulas in terms of the operators

$$\underset{s}{D}_i = \frac{\partial}{\partial x^i} + u_i^\alpha \frac{\partial}{\partial u^\alpha} + u_{ii_1}^\alpha \frac{\partial}{\partial u_{i_1}^\alpha} + \cdots + u_{ii_1\cdots i_s}^\alpha \frac{\partial}{\partial u_{i_1\cdots i_s}^\alpha}, \tag{10.9}$$

where s is a positive integer and $i=1,\cdots,n$. These operators naturally arise in group-theoretic calculations.

Now by virtue of (10.2) it is possible to express the differentials du^α, $\cdots, du_{i_1\cdots i_{k-1}}^\alpha$ in terms of the independent quantities $u_i^\alpha,\cdots,u_{i_1\cdots i_k}^\alpha, dx^i$. These expressions plus (10.6), when substituted into (10.8), yield—after rearrangement, use of the operator $\underset{s}{D}_i$, and independence of the dx^i and $du_{i_1\cdots i_k}^\alpha$—the following equivalent system of equations:

$$\begin{aligned} &\underset{k-1}{D}_i(\eta^\alpha) - u_j^\alpha \underset{k-1}{D}_i(\xi^j) - \zeta_i^\alpha = 0, \\ &\underset{k-1}{D}_i(\zeta_{i_1\cdots i_s}^\alpha) - u_{i_1\cdots i_s j}^\alpha \underset{k-1}{D}_i(\xi^j) - \zeta_{i_1\cdots i_s i}^\alpha = 0, \quad s=1,\cdots,k-1, \end{aligned} \tag{10.10}$$

and

$$(10.11)\qquad \begin{gathered}\frac{\partial \eta^\alpha}{\partial u^\beta_{i_1\cdots i_k}} - u^\alpha_i \frac{\partial \xi^i}{\partial u^\beta_{i_1\cdots i_k}} = 0,\\ \frac{\partial \zeta^\alpha_{i_1\cdots i_s}}{\partial u^\beta_{j_1\cdots j_k}} - u^\alpha_{i_1\cdots i_s j}\frac{\partial \xi^j}{\partial u^\beta_{j_1\cdots j_k}} = 0, \quad s=1,\cdots,k-1.\end{gathered}$$

Before attempting to solve these equations, we first rewrite (10.11) in terms of the new functions

$$(10.12)\qquad \begin{gathered} W^\alpha = \eta^\alpha - \xi^i u^\alpha_i,\\ W^\alpha_{i_1\cdots i_s} = \zeta^\alpha_{i_1\cdots i_s} - \xi^j u^\alpha_{i_1\cdots i_s j}, \quad s=1,\cdots,k-1,\end{gathered}$$

and obtain the equivalent system

$$(10.13)\qquad \begin{cases} \dfrac{\partial W^\alpha}{\partial u^\beta_{j_1\cdots j_k}} = 0,\\[2mm] \dfrac{\partial W^\alpha_{i_1\cdots i_s}}{\partial u^\beta_{j_1\cdots j_k}} = 0, \quad s=1,\cdots,k-2,\end{cases}$$

$$(10.14)\qquad \frac{\partial W^\alpha_{i_1\cdots i_{k-1}}}{\partial u^\beta_{j_1\cdots j_k}} = \xi^{j_n}\delta^\alpha_\beta \delta^{j_1\cdots j_{k-1}}_{i_1\cdots i_{k-1}}.$$

Let us now consider the case $m>1$; then (10.14) implies

$$(10.15)\qquad \frac{\partial W^\alpha_{i_1\cdots i_{k-1}}}{\partial u^\beta_{j_1\cdots j_k}} = 0, \quad \alpha\neq\beta,$$

for all values of the indices α, $i_1,\cdots,i_{k-1},j_1,\cdots,j_k$, and

$$(10.16)\qquad \xi^j = -\frac{\partial W^\alpha_{i_1\cdots i_{k-1}}}{\partial u^\alpha_{i_1\cdots i_{k-1}j}}$$

(no summation on any of the indices).

From (10.16) we have immediately that the ξ^j do not depend on $\underset{k}{u}$, and the general solution of (10.16) is

$$W^\alpha_{i_1\cdots i_{k-1}} = U^\alpha_{i_1\cdots i_{k-1}}\left(x, u, \underset{1}{u}, \cdots, \underset{k-1}{u}\right) - \xi^j\left(x, u, \underset{1}{u}, \cdots, \underset{k-1}{u}\right)u^\alpha_{i_1\cdots i_{k-1}j},$$

with arbitrary functions $U^\alpha_{i_1\cdots i_{k-1}}$, $\alpha=1,\cdots,m$. Further, from the definitions (10.12) and equations (10.13), we obtain that the coordinates ξ^i, η^α, $\zeta^\alpha_i,\cdots,\zeta^\alpha_{i_1\cdots i_{k-1}}$ of the operator (10.3) do not depend on $\underset{k}{u}$. Lie's theory of continuous groups gives us immediately that the transformation laws of the quantities $x, u, \underset{1}{u}, \cdots, \underset{k-1}{u}$ in (10.1) do not depend upon $\underset{k}{u}$. By induction and Theorem 9.1 we obtain that for $m>1$, $\xi^i=\xi^i(x, u)$, $\eta^\alpha=\eta^\alpha(x, u)$ and the coordinates $\zeta^\alpha_{i_1\cdots i_s}$ $(s=1,\cdots,k)$ are given by (10.10). This means that G given by (10.1) is the kth order extension of the group $\mathbb{P}$ of Lie point transformations in the space of (x, u) only.

Now we consider the case $m=1$. If $n>1$, we obtain from (10.14)

$$\xi^j=-\frac{\partial W_{i_1\cdots i_{k-1}}}{\partial u_{i_1\cdots i_{k-1}j}} \qquad \text{(no sum)},$$

$$\left.\frac{\partial W_{i_1\cdots i_{k-1}}}{\partial u_{j_1\cdots j_{k-1}j}}\right|_{(j_1,\cdots,j_{k-1})\neq(i_1,\cdots,i_{k-1})}=0.$$

From this, as in the preceding case, we have that the ξ^j's do not depend on $\underset{k}{u}$ and the general solution of the above equation is

$$W_{i_1\cdots i_{k-1}}=U_{i_1\cdots i_{k-1}}\left(x, u, \underset{1}{u}, \cdots, \underset{k-1}{u}\right)-\xi^j\left(x, u, \underset{1}{u}, \cdots, \underset{k-1}{u}\right)u_{i_1\cdots i_{k-1}j},$$

and again by induction, we prove that $\xi^i=\xi^i(x, u, \underset{1}{u})$, $\eta=\eta\,(x, u, \underset{1}{u})$, $\zeta_i=\zeta_i\,(x, u, \underset{1}{u})$, and the other coordinates $\zeta_{i_1\cdots i_s}$ $(s=2,\cdots,k)$ are given by extension formulas. This means that the group $\underset{k}{G}$ is the kth order extension of a group of Lie tangent transformations.

The same conclusion is valid for $m=n=1$ and can be proved in the following way. In this case (10.12)–(10.14) yield

$$\begin{aligned} W&=\eta-\xi\underset{1}{u},\\ W_1&=\zeta_1-\xi\underset{2}{u},\\ &\;\vdots\\ W_{k-2}&=\underset{k-2}{\zeta}-\xi\underset{k-1}{u},\\ W_{k-1}&=\underset{k-1}{\zeta}-\xi\underset{k}{u}, \end{aligned}\tag{10.17}$$

where

$$\zeta_s=\zeta_{\underbrace{1\cdots 1}_{s}},$$

and

$$\frac{\partial W}{\partial \underset{k}{u}}=0,\cdots,\quad \frac{\partial W_{k-2}}{\partial \underset{k}{u}}=0,\tag{10.18}$$

$$\xi=-\frac{\partial W_{k-2}}{\partial \underset{k}{u}}.\tag{10.19}$$

Using the last equation in (10.17) together with (10.10) we obtain

$$\underset{k-1}{D}(W_{k-2}) = \underset{k-1}{D}\left(\underset{k-2}{\zeta}\right) - \underset{k-1}{u}\,\underset{k-1}{D}(\xi) - \underset{k}{u}\xi$$

$$= \underset{k-1}{\zeta} - \xi\underset{k}{u} = W_{k-1},$$

or using the definition of $\underset{k-1}{D}$,

$$W_{k-1} = \frac{\partial W_{k-2}}{\partial x} + \cdots + \underset{k-1}{u}\frac{\partial W_{k-2}}{\partial \underset{k-2}{u}} + \underset{k}{u}\frac{\partial W_{k-2}}{\partial \underset{k-1}{u}}.$$

The latter equation together with (10.17) and (10.19) implies that

$$\xi = -\frac{\partial W_{k-2}}{\partial \underset{k-1}{u}},$$

$$\eta = W - \underset{1}{u}\frac{\partial W_{k-2}}{\partial \underset{k-1}{u}},$$

$$\underset{1}{\zeta} = W_1 - \underset{2}{u}\frac{\partial W_{k-2}}{\partial \underset{k-1}{u}},$$

$$\vdots$$

$$\underset{k-2}{\zeta} = W_{k-2} - \underset{k-1}{u}\frac{\partial W_{k-2}}{\partial \underset{k-1}{u}},$$

$$\underset{k-1}{\zeta} = \frac{\partial W_{k-2}}{\partial x} + \cdots + \underset{k-1}{u}\frac{\partial W_{k-2}}{\partial \underset{k-2}{u}}.$$

Equations (10.18) show that the coordinates $\xi, \eta, \cdots, \underset{k-1}{\zeta}$ depend only on x, $u, \underset{1}{u}, \cdots, \underset{k-1}{u}$. Hence these considerations show that there do not exist any finite-order tangent transformations beyond those considered by S. Lie. This statement can be formulated in the following way.

THEOREM 10.1: *Every group of the kth order tangent transformations is either*:

(i) *an extended group of Lie point transformations* $\mathbb{P}$ *in the case of more than one function, or*

(ii) *an extended group of Lie tangent transformations in the case of one function.*

II. INFINITE-ORDER TANGENT TRANSFORMATIONS

§11. Lie-Bäcklund Tangent Transformation Groups

The essence of the generalization discussed here is the wedding of Bäcklund's idea of infinite-order tangent transformations with Lie's notion of continuous groups of transformations. Let $x=(x^1,\cdots,x^n)\in\mathbb{R}^n, u=(u^1,\cdots,u^m)\in\mathbb{R}^m$, and for every $s=1, 2, 3,\cdots$ let $\underset{s}{u}$ be the set of quantities $u^\alpha_{i_1\cdots i_s}$ $(\alpha=1,\cdots,m,$ $i_s=1,\cdots,n)$ symmetric in their lower indices. Let us consider a one-parameter group G of point transformations

(11.1)
$$G:\begin{cases} x'^i=f^i\left(x, u, \underset{1}{u}, \underset{2}{u},\cdots; a\right),\\ u'^\alpha=\phi^\alpha\left(x, u, \underset{1}{u}, \underset{2}{u},\cdots; a\right),\\ u_i'^\alpha=\psi_i^\alpha\left(x, u, \underset{1}{u}, \underset{2}{u},\cdots; a\right),\\ \cdots\cdots\cdots\cdots\cdots \end{cases}$$

in the infinite-dimensional $(x, u, \underset{1}{u}, \underset{2}{u},\cdots)$-space. Together with the groups G, we consider its extension $\tilde{G}$ to the differentials dx^i, du^α, $du_i^\alpha,\cdots$ by means of the formulas

(11.2)
$$\begin{aligned} dx'^i&=\frac{\partial f^i}{\partial x^j}dx^j+\frac{\partial f^i}{\partial u^\beta}du^\beta+\frac{\partial f^i}{\partial u_j^\beta}du_j^\beta+\cdots,\\ du'^\alpha&=\frac{\partial \phi^\alpha}{\partial x^j}dx^j+\frac{\partial \phi^\alpha}{\partial u^\beta}du^\beta+\frac{\partial \phi^\alpha}{\partial u_j^\beta}du_j^\beta+\cdots,\\ du_i'^\alpha&=\frac{\partial \psi_i^\alpha}{\partial x^j}dx^j+\frac{\partial \psi_i^\alpha}{\partial u^\beta}du^\beta+\frac{\partial \psi_i^\alpha}{\partial u_j^\beta}du_j^\beta+\cdots\\ &\cdots\cdots\cdots\cdots\cdots \end{aligned}$$

DEFINITION 11.1. A group G is called a group of *Lie-Bäcklund tangent transformations* if the infinite system of equations

(11.3)
$$\begin{aligned} &du^\alpha-u_j^\alpha\,dx^j=0,\\ &du_i^\alpha-u_{ij}^\alpha\,dx^j=0,\\ &du_{i_1i_2}^\alpha-u_{i_1i_2j}^\alpha\,dx^j=0,\\ &\cdots\cdots\cdots\cdots \end{aligned}$$

is invariant with respect to the group $\tilde{G}$.

Now we state the infinitesimal invariance criteria of (11.3) in a compact form employing the differential operator

$$D_i = \frac{\partial}{\partial x^i} + u_i^\alpha \frac{\partial}{\partial u^\alpha} + u_{ii_1}^\alpha \frac{\partial}{\partial u_{i_1}^\alpha} + u_{ii_1 i_2}^\alpha \frac{\partial}{\partial u_{i_1 i_2}^\alpha} + \cdots, \tag{11.4}$$

which acts on functions of the independent variables $x^i, u^\alpha, u_i^\alpha, u_{ij}^\alpha, \cdots$.

Let

$$X = \xi^i \frac{\partial}{\partial x^i} + \eta^\alpha \frac{\partial}{\partial u^\alpha} + \zeta_i^\alpha \frac{\partial}{\partial u_i^\alpha} + \cdots + \zeta_{i_1 \cdots i_s}^\alpha \frac{\partial}{\partial u_{i_1 \cdots i_s}^\alpha} + \cdots \tag{11.5}$$

be the infinitesimal operator of the group G, where

$$\begin{gathered} \xi^i = \frac{\partial f^i}{\partial a}\bigg|_{a=0}, \qquad \eta^\alpha = \frac{\partial \phi^\alpha}{\partial a}\bigg|_{a=0}, \\ \zeta_{i_1 \cdots i_s}^\alpha = \frac{\partial \psi_{i_1 \cdots i_s}^\alpha}{\partial a}\bigg|_{a=0}, \quad s = 1, 2, 3, \cdots. \end{gathered} \tag{11.6}$$

In this case as in the finite-dimensional case, the infinitesimal operator (11.5) fully characterizes the group of transformations (11.1) if we insure the existence and uniqueness of the solution of the Lie-Bäcklund equation:

$$\frac{dF}{da} = \Xi(F), \qquad F|_{a=0} = z, \tag{11.7}$$

where

$$\begin{gathered} z = (x^i, u^\alpha, u_i^\alpha, \cdots), \\ F = (f^i, \phi^\alpha, \psi_i^\alpha, \cdots), \\ \Xi = (\xi^i, \eta^\alpha, \zeta_i^\alpha, \cdots). \end{gathered} \tag{11.8}$$

When the conditions of existence and uniqueness of the solution of the Lie-Bäcklund equation are satisfied, the group property of the transformations (11.1),

$$F(F(z, a), b) = F(z, a+b),$$

follows immediately from the uniqueness of the solution.

The infinitesimal operator $\tilde{X}$ of the group $\tilde{G}$ will have the form

$$\tilde{X} = X + \tilde{\xi}^i \frac{\partial}{\partial (dx^i)} + \tilde{\eta}^\alpha \frac{\partial}{\partial (du^\alpha)} + \tilde{\zeta}_i \frac{\partial}{\partial (du_i^\alpha)} + \cdots, \tag{11.9}$$

where

$$\tilde{\xi}^i = \frac{\partial (dx'^i)}{\partial a}\bigg|_{a=0}, \quad \tilde{\eta}^\alpha = \frac{\partial (du'^\alpha)}{\partial a}\bigg|_{a=0}, \quad \tilde{\zeta}_i^\alpha = \frac{\partial (du_i'^\alpha)}{\partial a}\bigg|_{a=0}, \cdots,$$

and from (11.2) and (11.6) we have

$$
\begin{aligned}
\tilde{\xi}^i &= \frac{\partial \xi^i}{\partial x^j} dx^j + \frac{\partial \xi^i}{\partial u^\beta} du^\beta + \frac{\partial \xi^i}{\partial u_j^\beta} du_j^\beta + \cdots, \\
\tilde{\eta}^\alpha &= \frac{\partial \eta^\alpha}{\partial x^j} dx^j + \frac{\partial \eta^\alpha}{\partial u^\beta} du^\beta + \frac{\partial \eta^\alpha}{\partial u_j^\beta} du_j^\beta + \cdots, \\
\tilde{\zeta}_i^\alpha &= \frac{\partial \zeta_i^\alpha}{\partial x^j} dx^j + \frac{\partial \zeta_i^\alpha}{\partial u^\beta} du^\beta + \frac{\partial \zeta_i^\alpha}{\partial u_j^\beta} du_j^\beta + \cdots, \\
&\cdots\cdots\cdots\cdots\cdots\cdots
\end{aligned}
\tag{11.10}
$$

The infinitesimal conditions for the invariance of system (11.3) are given by

$$
\begin{aligned}
\tilde{X}\left(du^\alpha - u_j^\alpha dx^j\right)\big|_{(11.3)} &= 0, \\
\tilde{X}\left(du_{i_1\cdots i_s}^\alpha - u_{i_1\cdots i_s j}^\alpha dx^j\right)\big|_{(11.3)} &= 0, \quad s = 1, 2, \cdots.
\end{aligned}
\tag{11.11}
$$

Immediately, we obtain from (11.9) and (11.10) the infinite analogue of (10.7), but in contrast to the case of finite-order transformations, we obtain only the infinite analogue of equations (10.9) as the equivalent form of equations (11.11), because the only independent differentials now are the $dx^i (i = 1, \cdots, n)$. As a result we obtain the following theorem for the infinitesimal characterization of groups of Lie-Bäcklund tangent transformations.

THEOREM 11.1 *The group G of tranformations* (11.1) *is a group of Lie-Bäcklund tangent transformations if and only if the coordinates of the infinitesimal operator* (11.5) *satisfy the following equations*:

$$
\begin{aligned}
\zeta_i^\alpha &= D_i(\eta^\alpha) - u_j^\alpha D_i(\xi^j), \\
\zeta_{i_1 i_2}^\alpha &= D_{i_2}(\zeta_{i_1}^\alpha) - u_{i_1 j}^\alpha D_{i_2}(\xi^j), \\
\zeta_{i_1\cdots i_s}^\alpha &= D_{i_s}(\zeta_{i_1\cdots i_{s-1}}^\alpha) - u_{i_1\cdots i_{s-1} j}^\alpha D_{i_s}(\xi^j), \\
&\cdots\cdots\cdots\cdots\cdots\cdots
\end{aligned}
\tag{11.12}
$$

Remark. If, in the transformations (11.1), the transformed quantities x' and u' depend only upon x, u, and the group parameter a, then Theorem 11.1 shows us that the group G of Lie-Bäcklund transformations must be the infinite extension of a group $\mathbb{P}$ of Lie point transformations (9.5). In this case equations (11.12) become the well-known extension formulas.

Introducing the quantities

$$\mu^\alpha = \eta^\alpha - u_j^\alpha \xi^j,$$

(11.12) can be rewritten as

$$
\begin{aligned}
&\zeta_i^\alpha = D_i(\mu^\alpha) + \xi^j u_{ij}^\alpha, \\
&\zeta_{i_1 i_2}^\alpha = D_{i_1} D_{i_2}(\mu^\alpha) + \xi^j u_{i_1 i_2 j}^\alpha, \\
&\vdots \\
&\zeta_{i_1 \cdots i_s}^\alpha = D_{i_1} \cdots D_{i_s}(\mu^\alpha) + \xi^j u_{i_1 \cdots i_s j}^\alpha, \\
&\cdots\cdots\cdots\cdots\cdots\cdots
\end{aligned}
\tag{11.13}
$$

The operator (11.5) with coordinates $\zeta_{i_1 \cdots i_s}^\alpha$ defined by (11.13) is called a Lie-Bäcklund operator and will be denoted frequently by

$$
X = \xi^i \frac{\partial}{\partial x^i} + \eta^\alpha \frac{\partial}{\partial u^\alpha} + \cdots. \tag{11.14}
$$

Generalizing the considerations of Bäcklund [2], we now consider whether it is possible in this formulation of tangent transformations to find a finite-dimensional space of the variables $x, u, \underset{1}{u}, \cdots, \underset{k}{u}$ which transforms into itself under the transformations (11.1). Therefore we assume that the transformations (11.1) take the form

$$
\begin{aligned}
&x'^i = f^i(x, u, \underset{1}{u}, \cdots, \underset{k}{u}; a), \\
&u'^\alpha = \phi^\alpha(x, u, \underset{1}{u}, \cdots, \underset{k}{u}; a), \\
&u'^\alpha = \psi_i^\alpha(x, u, \underset{1}{u}, \cdots, \underset{k}{u}; a), \\
&\vdots \\
&u'^\alpha_{i_1 \cdots i_k} = \psi_{i_1 \cdots i_k}^\alpha(x, u, \underset{1}{u}, \cdots, \underset{k}{u}; a), \\
&u'^\alpha_{i_1 \cdots i_{k+1}} = \psi_{i_1 \cdots i_{k+1}}^\alpha(x, u, \underset{1}{u}, \cdots; a), \\
&\cdots\cdots\cdots\cdots\cdots\cdots
\end{aligned}
\tag{11.15}
$$

In this case the coefficients ξ^i, η^α, $\zeta_{i_1 \cdots i_s}^\alpha (s = 1, \cdots, k)$ depend only on x, $u, \underset{1}{u}, \cdots, \underset{k}{u}$.

First we take the case $k = 1$. Let's consider the first of equations (11.12):

$$
\zeta_i^\alpha = D_i(\eta^\alpha) - u_j^\alpha D_i(\xi^j).
$$

Because ζ_i depends only on x, u, $\underset{1}{u}$ and

$$D_i(\eta^\alpha)-u_j^\alpha D_i(\xi^j)=\frac{\partial\eta^\alpha}{\partial x^i}+u_i^\beta\frac{\partial\eta^\alpha}{\partial u^\beta}$$

$$-u_j^\alpha\left(\frac{\partial\xi^j}{\partial x^i}+u_i^\beta\frac{\partial\xi^j}{\partial u^\beta}\right)+u_{ik}^\beta\left(\frac{\partial\eta^\alpha}{\partial u_k^\beta}-u_j^\alpha\frac{\partial\xi^j}{\partial u_k^\beta}\right),$$

we immediately obtain equations (9.9). As a result we find that G is the extension of a group of Lie tangent contact transformations in the case of one function, and G is the extension of the group $\mathbb{P}$ of Lie point transformations (9.5) in the case of more than one function.

Now let's take the case $k>1$. In this case we consider the first n equations of (11.12):

$$\zeta_i^\alpha=D_i(\eta^\alpha)-u_j^\alpha D_j(\xi^j),$$

$$\zeta_{i\cdots i_{s+1}}^\alpha=D_{i_{s+1}}(\zeta_{i_1\cdots i_s}^\alpha)-u_{i_1\cdots i_s j}^\alpha D_{i_{s+1}}(\xi^j),\quad s=1,\cdots,k-1.$$

The right-hand sides of these equations can be rewritten by using the definition (1.21) of the operator $\underset{k-1}{D}$ as

$$D_i(\eta^\alpha)-u_j^\alpha D_i(\xi^j)=\underset{k-1}{D}{}_i(\eta^\alpha)-u_j^\alpha\underset{k-1}{D}{}_i(\xi^j)$$

$$+u_{ii_1\cdots i_k}^\beta\left(\frac{\partial\eta^\alpha}{\partial u_{i_1\cdots i_k}^\beta}-u_j^\alpha\frac{\partial\xi^j}{u_{i_1\cdots i_k}^\beta}\right),$$

$$D_{i_{s+1}}(\zeta_{i_1\cdots i_s}^\alpha)-u_{i_1\cdots i_s j}^\alpha D_{i_{s+1}}(\xi^j)=\underset{k-1}{D}{}_{i_{s+1}}(\zeta_{i_1\cdots i_s}^\alpha)-u_{i_1\cdots i_s j}^\alpha\underset{k-1}{D}{}_{i_{s+1}}(\xi^j)$$

$$+u_{i_{s+1}j_1\cdots j_k}^\beta\left(\frac{\partial\zeta_{i_1\cdots i_s}^\alpha}{\partial u_{j_1\cdots j_k}^\beta}-u_{i_1\cdots i_s j}^\alpha\frac{\partial\xi^j}{\partial u_{j_1\cdots j_k}^\beta}\right),$$

where the terms involving $\underset{k-1}{D}$ are independent of the coordinates of $\underset{k+1}{u}$. Now because the left-hand sides $\zeta_{i_1\cdots i_s}^\alpha$ $(s=1,\cdots,k)$ of the same equations depend only on x, u, $\underset{1}{u},\cdots,\underset{k}{u}$, we immediately obtain (10.9) and (10.10). Therefore using the same argument as in §10, we obtain that in this case the coordinates of the operator (11.5) depend only on x, u. So besides groups $\mathbb{P}$ of Lie point transformations and the groups of Lie tangent transformations there exist no groups of Lie-Bäcklund tangent transformations which act invariantly on a finite-dimensional $(x, u, \underset{1}{u},\cdots,\underset{k}{u})$-space for any $k\geqq 1$. Hence the theory of groups of Lie-Bäcklund tangent transformations is essentially a transformation theory of an infinite-dimensional space.

§12. Lie-Bäcklund Equations[1]

For a given Lie-Bäcklund operator

$$X=\xi^i\Big(x,u,\underset{1}{u},\cdots\Big)\frac{\partial}{\partial x^i}+\eta^\alpha\Big(x,u,\underset{1}{u},\cdots\Big)\frac{\partial}{\partial u^\alpha}$$
$$+\zeta_i^\alpha\Big(x,u,\underset{1}{u},\cdots\Big)\frac{\partial}{\partial u_i^\alpha}+\zeta_{i_1i_2}^\alpha\Big(x,u,\underset{1}{u},\cdots\Big)\frac{\partial}{\partial u_{i_1i_2}^\alpha}+\cdots, \tag{12.1}$$

where

$$\begin{aligned}&\zeta_i^\alpha=D_i(\eta^\alpha-u_j^\alpha\xi^j)+\xi^ju_{ij}^\alpha,\\ &\zeta_{i_1i_2}^\alpha=D_{i_1}D_{i_2}(\eta^\alpha-u_j^\alpha\xi^j)+\xi^ju_{i_1i_2j}^\alpha,\\ &\cdots\cdots\cdots\cdots\cdots\cdots,\end{aligned} \tag{12.2}$$

there exists a one-parameter group of Lie-Bäcklund transformations if the solution of the Lie-Bäcklund equations

$$\begin{aligned}&\frac{dx'^i}{da}=\xi^i(x',u',\underset{1}{u}',\cdots),\\ &\frac{du'^\alpha}{da}=\eta^\alpha(x',u',\underset{1}{u}',\cdots),\\ &\frac{du_i'^\alpha}{da}=\zeta_i^\alpha(x',u',\underset{1}{u}',\cdots),\\ &\cdots\cdots\cdots\cdots\cdots,\end{aligned} \tag{12.3}$$

with initial data

$$x'^i|_{a=0}=x^i,\quad u'^\alpha|_{a=0}=u^\alpha,\quad u_i'^\alpha|_{a=0}=u_i^\alpha,\cdots, \tag{12.4}$$

exists and is unique. This group property follows directly from uniqueness, as in the theory of Lie tangent transformations. Therefore the basic problem here is the integrability of the Lie-Bäcklund equation (12.3), which constitutes an autonomous system of ordinary differential equations in an infinite-dimensional space. This question is open in general, and here we confine our discussion to some examples of integrable Lie-Bäcklund equations.

Well-known classes of examples are provided by Lie tangent transformation groups in the case of one dependent variable u, and Lie point transformations in the case of several dependent functions $u=(u^1,\cdots,u^m)$.

There exist other examples which represent realizations of integrable Lie-Bäcklund equations beyond those of Lie. In the following examples we shall employ the notation of §11, in particular (11.7) and (11.8).

[1]See Ibragimov [8].

Example 1. Consider a Banach space B of elements $z=(x, u, \underset{1}{u}, \underset{2}{u},\cdots)$ with norm

$$\|z\|=|x|+\sum_{k=0}^{\infty}\left|\underset{k}{u}\right|<\infty,$$

where $|x|$ is any norm in the finite-dimensional Banach space of elements $x=(x^1,\cdots,x^n)$, and similarly for $\left|\underset{k}{u}\right|$. For simplicity we restrict ourselves to the case $n=1$, $m=1$. Now take in this space the Lie-Bäcklund equation

$$\frac{dz'}{da}=\Xi(z'), \qquad z'|_{a=0}=z, \tag{12.5}$$

with

$$\Xi(z)=(0, \eta(z), D\eta(z),\cdots), \tag{12.6}$$

where

$$\eta^{\alpha}(z)=\sum_{i=0}^{\nu}\alpha_i \underset{i}{u},$$

with $\alpha_i=\text{const}$, and ν any positive integer. It then follows that

$$\|\Xi(z)-\Xi(z')\|=C\|z-z'\| \quad \text{for } C<\infty,$$

and according to the theory of differential equations in Banach spaces, this estimate insures the existence and uniqueness of the solution of the Lie-Bäcklund equation (12.5) prescribed by (12.6).

Example 2. Take $\eta^{\alpha}=\xi^i u_i^{\alpha}$, $\alpha=1,\cdots,m$, in (12.5), and let the ξ^i be arbitrary smooth functions depending on a finite number of "derivatives," that is, $\xi^i=\xi^i(x, u, \underset{1}{u},\cdots, \underset{s}{u})$. Here Ovsjannikov's theorem (Ovsjannikov [3]) is applicable. As a basic space we'll choose the space of elements $z=(x, u, \underset{1}{u},\cdots)$ defined in the following manner. Let

$$\|x\|=\max_i|x^i|, \qquad \|u\|=\max_{\alpha}|u^{\alpha}|,$$

$$\left\|\underset{s}{u}\right\|=\max_{\alpha, i_1,\cdots,i_s}|u^{\alpha}_{i_1\cdots i_s}|, \qquad \left\|\xi^i(z)\underset{s+1}{u}{}_i\right\|=\max_{\alpha, i_1,\cdots,i_s}\left|\xi^i(z)u^{\alpha}_{ii_1\cdots i_s}\right|.$$

For any real $\rho>0$ the norm

$$\|z\|_{\rho}=\|x\|+\sum_{s=0}^{\infty}\frac{\rho^s}{s!}\left\|\underset{s}{u}\right\|$$

is determined, so that z is an element of a Banach space B_{ρ}.

Then Ovsjannikov's theorem insures the existence and uniqueness of the solution of the equation (12.5) if the mapping Ξ is quasidifferential, that is, for points z, z' of some small sphere the inequality

$$\|\Xi(z)-\Xi(z')\|_\rho \leqq \left(1+\frac{\partial}{\partial\rho}\right)\{F(\|z\|_\rho+\|z'\|_\rho)\cdot\|z-z'\|_\rho\}$$

is valid, where F is a real function satisfying the following conditions:

$$F(\tau)\geq 0,\quad \frac{dF(\tau)}{d\tau}\geq 0,\quad \frac{d^2F(\tau)}{d\tau^2}\geq 0,\qquad \text{for } \tau\geq 0.$$

In our case, we have

$$\Xi(z)=(\xi^i(z), \xi^j(z)u_j^\alpha, \xi^j(z)u_{ij}^\alpha,\cdots)$$

and

$$\|\Xi(z)-\Xi(z')\|_\rho=\|\xi(z)-\xi(z')\|+\sum_{s=0}^{\infty}\frac{\rho^s}{s!}\left\|\xi^i(z)\underset{s+1}{u}{}_i-\xi^i(z')\underset{s+1}{u'}{}_i\right\|.$$

The estimate for this difference is presented elsewhere (Ibragimov [7]) and is given by

$$\|\Xi(z)-\Xi(z')\|_\rho \leqslant \left(1+\frac{\partial}{\partial\rho}\right)\{C\cdot(1+\|z\|_\rho+\|z'\|_\rho)\cdot\|z-z'\|_\rho\},\qquad C=\text{const}.$$

Hence the operation considered is quasidifferential, and the existence and uniqueness of the corresponding Lie-Bäcklund equation is insured.

Further examples are obtained by noting that if

$$(12.7)\qquad \frac{dx'^i}{da}=0,\qquad \frac{du'^\alpha}{da}=\eta^\alpha\left(x, u', \underset{1}{u'},\cdots\right),$$

it then follows that

$$\frac{du_i'^\alpha}{da}=D_i\zeta_i^\alpha=D_i\eta^\alpha=\frac{du'^\alpha}{da}.$$

Therefore, as an alternative method to directly integrating (12.3) with initial data (12.4), we can consider (12.7) as the partial differential equation (Anderson, Barut, and Ibragimov [1], and §16)

$$(12.8)\qquad \frac{\partial u'^\alpha}{\partial a}=\eta^\alpha\left(x, u', \frac{\partial u'}{\partial x},\cdots\right),$$

with Cauchy data

$$(12.9)\qquad u'|_{a=0}=u(x).$$

In particular, consider the following two examples:

Example 3. In (12.7) take the case $x \in \mathbb{R}$, $u \in \mathbb{R}$ and $\eta = \underset{1}{u}$, which implies that (12.8) becomes

$$\frac{\partial u'}{\partial a} = \frac{\partial u'}{\partial x}, \tag{12.10}$$

with the Cauchy data

$$u'|_{a=0} = u\,(x). \tag{12.11}$$

Suppose the function $u\,(x)$ is an analytical one; then we find the solution of the Cauchy problem (12.10), (12.11) to be

$$u'(x, a) = u\,(x+a).$$

Hence the corresponding Lie-Bäcklund equation is a group of translations. This implies via a Taylor series expansion that the corresponding point form is

$$u' = u + \sum_{n=1}^{\infty} \frac{a^n}{n!} \underset{n}{u}.$$

Later we'll see that this result is connected with the equivalence structure for Lie-Bäcklund transformation groups.

Example 4. Similarly, for the case $\eta = \underset{2}{u}$, we have that the corresponding Lie-Bäcklund equation (12.5) is equivalent to the Cauchy problem for the heat equation:

$$\frac{\partial u'}{\partial a} = \frac{\partial^2 u'}{\partial x^2}, \qquad u'|_{a=0} = u\,(x). \tag{12.12}$$

Taking the classical solution

$$u'(x, a) = \frac{1}{\sqrt{\pi}} \int_{-\infty}^{x} u\,(x + 2\sqrt{a}\, b) e^{-b^2} db, \quad x > 0,$$

in the Jevrey class $C^{1/2}$ (Hadamard [1]), then the corresponding point form (again obtained via a Taylor series expansion) is

$$u' = \sum_{n=0}^{\infty} \frac{a^n}{n!} \underset{2n}{u},$$

$$\underset{2s}{u'} = \sum_{n=0}^{\infty} \frac{a^n}{n!} \underset{2(n+s)}{u}, \quad s = 1, 2, \cdots,$$

$$\underset{2s+1}{u'} = \sum_{n=0}^{\infty} \frac{a^n}{n!} \underset{2(n+s)+1}{u}, \quad s = 0, 1, \cdots.$$

The second example in the preceding discussion of integrability leads to the result that if two Lie-Bäcklund operators (11.14) differ by an operator of the type

$$X_0=\xi_0^i D_i=\xi_0^i \frac{\partial}{\partial x^i}+\xi_0^i u_i^k \frac{\partial}{\partial u^k}+\cdots, \tag{12.13}$$

where the functions ξ_0^i satisfy the integrability conditions for the corresponding Lie-Bäcklund equations (11.7), (11.8) (for instance, as in Example 2, it can be a smooth function on a finite number of derivatives, but it should be noted that many possibilities exist), then the integrability of the Lie-Bäcklund equations (11.7), (11.8) for one such operator implies the integrability of the Lie-Bäcklund equations for the other one. This situation leads to the consideration of equivalence classes of Lie-Bäcklund operators defined in the following manner. Consider two Lie-Bäcklund operators (11.14):

$$X=\xi^i \frac{\partial}{\partial x^i}+\eta^k \frac{\partial}{\partial u^k}+\cdots, \tag{12.14}$$

$$X_*=\xi_*^i \frac{\partial}{\partial x^i}+\eta_*^k \frac{\partial}{\partial u^k}+\cdots. \tag{12.15}$$

Definition 12.1. Any operator X_0 of the form (12.13) is called equivalent to 0, and we write $X_0 \sim 0$.

Definition 12.2. An operator X is equivalent to X_*, and we write $X \sim X_*$, if $X-X_* \sim 0$.

Now employing the results from Example 2 and Definitions 12.1, 12.2, it can be shown that the integrability of one member of an equivalence class implies the integrability of all members of that class.

Another consequence of these considerations appears in §13 in the discussion of the application of these transformations to differential equations, namely, the invariance of a given system of differential equations with respect to one Lie-Bäcklund operator implies the invariance of that system with respect to all the members of the equivalence class corresponding to that Lie-Bäcklund operator.

A further consequence of this structure is that we can establish the equivalence of a special type of Lie-Bäcklund transformation group to Lie tangent transformation groups. This result is formulated as the following theorem.

Theorem 12.1. *Any Lie-Bäcklund group of transformations with a Lie-Bäcklund operator* (11.14) *of the form*

$$X=\xi^i\left(x, u, \underset{1}{u}\right)\frac{\partial}{\partial x^i}+\eta\left(x, u, \underset{1}{u}\right)\frac{\partial}{\partial u}+\cdots$$

is equivalent to a group of Lie tangent transformations, i.e., $X \sim Y$, *where*

$$Y=-\frac{\partial W}{\partial u_i}\frac{\partial}{\partial x^i}+\left(W-u_i\frac{\partial W}{\partial u^i}\right)\frac{\partial}{\partial u}+\left(\frac{\partial W}{\partial x^i}+u_i\frac{\partial W}{\partial u}\right)\frac{\partial}{\partial u_i},$$

and W *is some function of* $x, u, \underset{1}{u}$ *only.*

Proof. Let the coefficients ξ^i, η of the Lie-Bäcklund operator

$$X=\xi^i\frac{\partial}{\partial x^i}+\eta\frac{\partial}{\partial u}+\cdots \tag{12.16}$$

be functions of x, u, $\underset{1}{u}$ only, then the operator (12.16) is equivalent to an operator Y corresponding to a Lie tangent transformation group:

$$Y=\tilde{\xi}^i\frac{\partial}{\partial x^i}+\tilde{\eta}\frac{\partial}{\partial u}+\tilde{\zeta}_i\frac{\partial}{\partial u_i},$$

with

$$\tilde{\xi}^i=-\frac{\partial W}{\partial u_i},\quad \tilde{\eta}=W-u_i\frac{\partial W}{\partial u_i},\quad \tilde{\zeta}_i=\frac{\partial W}{\partial x^i}+u_i\frac{\partial W}{\partial u},$$

where $W=\eta-u_j\xi^j$. Indeed,

$$\xi^i=-\frac{\partial W}{\partial u_i}+\xi^i+\frac{\partial W}{\partial u_i},$$

$$\eta=W-u_j\frac{\partial W}{\partial u_j}+u_j\left(\xi^j+\frac{\partial W}{\partial u_j}\right),$$

$$\zeta_i=\frac{\partial W}{\partial x^i}+u_i\frac{\partial W}{\partial u}+u_{ij}\left(\xi^j+\frac{\partial W}{\partial u_j}\right),$$

and therefore

$$X-Y=\left(\xi^i+\frac{\partial W}{\partial u_i}\right)D_i,$$

which is of the form (12.13). ∎

To illustrate the significance of the equivalence structure for Lie-Bäcklund transformation groups, consider the following example. The second-order ordinary differential equation

$$\frac{d^2u}{dx^2}=0 \tag{12.17}$$

admits a maximal 8-parameter point invariance transformation group G_8 (Lie [6], p. 298; also see §15), while the equivalent system

$$\frac{du^1}{dx}=u^2,\qquad \frac{du^2}{dx}=0 \tag{12.18}$$

admits an infinite invariance group G_∞. The equivalence structure clarifies the nature of this difference in the following manner. The most general invariance group of point transformations admitted by the system (12.18) is generated by an operator of the form

$$X=\xi\frac{\partial}{\partial x}+\eta^1\frac{\partial}{\partial u^1}+\eta^2\frac{\partial}{\partial u^2}, \tag{12.19}$$

where ξ, η^1, η^2 are functions of x, u^1, u^2 only, and can be reduced to a group generated by an operator of the type (12.19) with $\xi\equiv 0$. Here the invariance of the system (12.18) with respect to the group generated by

$$X_0=\xi\frac{\partial}{\partial x}+\xi u^2\frac{\partial}{\partial u^1} \tag{12.20}$$

is employed (Ovsjannikov [2]) to effect the reduction. Now, if an operator

$$X=\eta^1\frac{\partial}{\partial u^1}+\eta^2\frac{\partial}{\partial u^2} \tag{12.21}$$

is admissible by the system (12.18), then it readily follows that the most general form of (12.21) is described by coordinates

$$\begin{aligned}\eta^1&=u^1 g\,(u^1-xu^2, u^2)+h\,(u^1-xu^2, u^2),\\ \eta^2&=u^2 g\,(u^1-xu^2, u^2),\end{aligned} \tag{12.22}$$

where g, h are arbitrary functions. Rewriting these operators for the system (12.17), we obtain the following Lie-Bäcklund invariance operators for (12.17): the operator (12.20) becomes an operator of the type (12.13), and the operator (12.21) with (12.22) becomes the Lie-Bäcklund operator

$$X=\left[ug\left(u-x\underset{1}{u}, \underset{1}{u}\right)+h\left(u-x\underset{1}{u}, \underset{1}{u}\right)\right]\frac{\partial}{\partial u}+\cdots. \tag{12.23}$$

It can be directly established that the operator (12.23) is not the tangent vector of any Lie tangent transformation group. But it is equivalent to a Lie tangent operator

$$Y=-\frac{\partial W}{\partial \underset{1}{u}}\frac{\partial}{\partial x}+\left(W-\underset{1}{u}\frac{\partial W}{\partial \underset{1}{u}}\right)\frac{\partial}{\partial u}+\left(\frac{\partial W}{\partial x}+\underset{1}{u}\frac{\partial W}{\partial \underset{1}{u}}\right)\frac{\partial}{\partial \underset{1}{u}}, \tag{12.24}$$

where

$$W=ug\left(u-x\underset{1}{u}, \underset{1}{u}\right)+h\left(u-x\underset{1}{u}, \underset{1}{u}\right). \tag{12.25}$$

Now by direct calculation, it can be shown that Y given by (12.24) is the most general infinitesimal operator for a Lie tangent transformation group admitted by (12.17). Indeed, take the most general Lie tangent operator admitted by (12.17),

$$Y=\xi\left(x, u, \underset{1}{u}\right)\frac{\partial}{\partial x}+\eta\left(x, u, \underset{1}{u}\right)\frac{\partial}{\partial u}+\underset{1}{\zeta}\left(x, u, \underset{1}{u}\right)\frac{\partial}{\partial \underset{1}{u}}, \tag{12.26}$$

where

$$\xi=-\frac{\partial W}{\partial \underset{1}{u}},\quad \eta=W-\underset{1}{u}\frac{\partial W}{\partial \underset{1}{u}},\quad \zeta_1=\frac{\partial W}{\partial x}+\underset{1}{u}\frac{\partial W}{\partial x},\tag{12.27}$$

and its prolongation

$$\tilde{X}=X+\zeta_2\frac{\partial}{\partial \underset{2}{u}},\tag{12.28}$$

where (12.27) implies

$$\zeta_2=D_x D_x(W)+\xi\underset{2}{u}.\tag{12.29}$$

The defining equation which characterizes the invariance of (12.17) with respect to the Lie tangent transformation group generated by the infinitesimal operator (12.28) is

$$\zeta_2|_{(12.17)}=0,\tag{12.30}$$

with ζ_2 given by (12.29) or

$$\left(\frac{\partial}{\partial x}+\underset{1}{u}\frac{\partial}{\partial u}\right)^2 W=0.\tag{12.31}$$

Now take

$$V=\frac{\partial W}{\partial x}+\underset{1}{u}\frac{\partial W}{\partial u};\tag{12.32}$$

then (12.31) implies that V satisfies

$$\frac{\partial V}{\partial x}+\underset{1}{u}\frac{\partial V}{\partial u}=0\tag{12.33}$$

with general solution

$$V=\underset{1}{u}\,g\left(u-x\underset{1}{u},\underset{1}{u}\right),\tag{12.34}$$

where g is an arbitrary function. Substitution of the solution (12.34) into (12.32) yields

$$\frac{\partial W}{\partial x}+\underset{1}{u}\frac{\partial W}{\partial u}=\underset{1}{u}\,g\left(u-x\underset{1}{u},\underset{1}{u}\right),\tag{12.35}$$

with a particular solution W_p:

$$W_p=ug\left(u-x\underset{1}{u},\underset{1}{u}\right),$$

so that the general solution of (12.35) and consequently (12.31) is given by

$$W=ug\left(u-x\underset{1}{u},\underset{1}{u}\right)+h\left(u-x\underset{1}{u},\underset{1}{u}\right).\tag{12.36}$$

Hence (12.24) with W given by (12.25) is the most general form for an infinitesimal operator admitted by (12.17).

Chapter 3

Application to Differential Equations

In this chapter we trace a few applications of Lie-Bäcklund transformations to differential equations. These applications are chosen to hint at the potentialities inherent in the extension of Lie tangent transformations to Lie-Bäcklund transformations.

§13. Defining Equations

Consider a given system of differential equations

$$\omega : \omega_\nu\big(x, u, \underset{1}{u}, \cdots, \underset{k}{u}\big) = 0, \qquad \nu = 1, \cdots, M, \tag{13.1}$$

where $x = (x^1, \cdots, x^n) \in \mathbb{R}^n$, $u = (u^1, \cdots, u^m) \in \mathbb{R}^m$, and $\underset{1}{u}, \cdots, \underset{k}{u}$ are the corresponding 1st, $\cdots$, kth derivatives, respectively. Suppose that a Lie-Bäcklund operator

$$X = \xi^i\big(x, u, \underset{1}{u} \cdots\big)\frac{\partial}{\partial x^i} + \eta^\alpha\big(x, u, \underset{1}{u}, \cdots\big)\frac{\partial}{\partial u^\alpha} + \cdots \tag{13.2}$$

generates a Lie-Bäcklund transformation group G, i.e., the operator (13.2) is by assumption integrable to the corresponding group G. Now, if we act by the group G on the system ω given by (13.1), in general the order of these equations will increase, so that after the action of G we shall have in general a system of infinite-order partial differential equations. Therefore, if we want to define the invariance of ω with respect to the Lie-Bäcklund group G in the classical manner, as e.g. Lie point transformation groups, then we shall have a very strong requirement on the group G. Rather, it's natural to define the invariance of the system ω in a weaker manner, taking into account the differential consequences of (13.1), so that from the very beginning we have a system of equations in the infinite-dimensional space of variables $x, u, \underset{1}{u}, \cdots$, and then we require that G conserve this infinite system of differential equations. More precisely, this notion of invariance is given by the following definition.

DEFINITION 13.1. The system of differential equations (13.1) is called invariant with respect to a Lie-Bäcklund transformation group G if the manifold given by the following infinite system of differential equations

$$\left.\begin{aligned} \omega_\nu&=0,\\ D_i\omega_\nu&=0,\\ D_iD_j\omega_\nu&=0,\\ \cdots&\cdots \end{aligned}\right\}\qquad \nu=1,\cdots,M, \tag{13.3}$$

is an invariant manifold of the group G. This definition becomes the usual definition in the case of Lie point transformations, as follows from the fact that if the system (13.1) is invariant with respect to a group of Lie point transformations, then the manifold (13.3) is automatically invariant with respect to the infinite-order prolongation of this group of Lie point transformations.

In the case of Lie-Bäcklund transformation groups the infinite-order prolongation (13.3) of the system (13.1) is essential and nontrivial. Now, employing the same procedure as in the classical case of point transformations (Ovsjannikov [2]), we are led to the following criteria for the group invariance of (13.1):

THEOREM 13.1. *The system of differential equations* (13.1) *is invariant with respect to a Lie-Bäcklund transformation group G generated by a Lie-Bäcklund operator of the form* (13.2) *if and only if*

$$X\omega_\nu|_{(13.3)}=0,\qquad \nu=1,\cdots,M. \tag{13.4}$$

Proof. It follows directly from Definition 13.1 that the infinitesimal criterion for the invariance of the manifold given by (13.3) with respect to a Lie-Bäcklund group generated by a Lie-Bäcklund operator (13.2) is

$$\begin{aligned} X\omega_\nu|_{(13.3)}&=0,\\ XD_i\omega_\nu|_{(13.3)}&=0,\\ XD_iD_j\omega_\nu|_{(13.3)}&=0,\\ \cdots&\cdots \end{aligned} \tag{13.5}$$

The proof is completed by the observation that all the equations in (13.5) beyond the first are consequences of (13.4) because of the following identity:

$$D_iX-XD_i=D_i(\xi^j)D_j,\qquad i=1,\cdots,n, \tag{13.6}$$

which holds for any Lie-Bäcklund operator X (Ibragimov [6]). ■

Equation (13.4) can be called, as in the classical case, the *defining equation* for the Lie-Bäcklund group admitted by the system of differential equations (13.1). In general, the defining equation (13.4) is not of the same character as in the classical case, because (13.4) is no longer a system of partial differential equations for the coordinates ξ^i, η^α of the unknown Lie-Bäcklund operator X. Indeed, in general the coordinates ξ^i, η^α depend on derivatives which are not free, but which are connected by equations (13.3). As a result of this dependence, the defining equation (13.4) is an equation which relates the values of ξ^i, η^α and their derivatives at points $(x, u, \underset{1}{u}, \cdots)$ connected by (13.3), so that (13.4) is a system of differential functional equations. In the classical case of point transformations, the unknowns ξ^i, η^α are functions only of x, u and are not connected by the differential equations (13.1) or their differential consequences, so that in this case the defining equation is a differential equation. But this difficulty can be easily overcome. Let us divide the derivatives into two classes, "independent" and "dependent," in such a way that the "independent" ones completely specify the "dependent" ones according to (13.3). Then without loss of generality we can take the coordinates ξ^i, η^α of the Lie-Bäcklund operator under consideration to depend only on the x, u and the "independent" derivatives (see e.g., Ibragimov [6]).

Note that any operator (12.13) is admissible by any system of differential equations; therefore (13.1) admits together with any Lie-Bäcklund group G all the members of the same equivalence class. This permits us to further simplify the form of the operator (13.2) by choosing the "simplest" representative of a class of equivalent Lie-Bäcklund operators. One can set e.g. $\xi^i=0$ in (13.2). For this latter choice, the defining equation becomes the following system of differential equations for the unknown functions ξ^i:

$$\left.\left(\eta^\alpha\frac{\partial\omega_\nu}{\partial u^\alpha}+(D_i\eta^\alpha)\frac{\partial\omega_\nu}{\partial u_i^\alpha}+(D_{i_1}D_{i_2}\eta^\alpha)\frac{\partial\omega_\nu}{\partial u_{i_1i_2}^\alpha}+\cdots\right)\right|_{(13.3)}=0. \tag{13.7}$$

This consideration of invariance of a system of differential equations permits a generalization which is connected with the notion of Bäcklund transformations (see §6). Namely, given a system of differential equations (13.1), the problem is to find a transformation group (3.1) which is a group of tangent transformations on the solutions of (13.1). More precisely, find a transformation group (11.1) such that the joint system of equations (13.1) and (11.3) is invariant with respect to the transformations (11.1) for the manifold specified by (13.3). It follows from the preceding and Theorem 11.1 that the infinitesimal criterion for this invariance is given by

$$\begin{gathered}\left.\left(\zeta_i^\alpha-D_i(\eta^\alpha-\xi^ju_j^\alpha)-\xi^ju_{ij}^\alpha\right)\right|_{(13.3)}=0,\\ \left.\left(\zeta_{i_1i_2}^\alpha-D_{i_1}D_{i_2}(\eta^\alpha-\xi^ju_j^\alpha)-\xi^ju_{i_1i_2j}^\alpha\right)\right|_{(13.3)}=0,\\ \cdots\cdots\cdots\cdots\cdots\cdots\cdots\cdots\\ X\omega_\nu|_{(13.3)}=0,\quad XD_i\omega_\nu|_{(13.3)}=0,\cdots.\end{gathered} \tag{13.8}$$

We can analyze the solutions of a given system of differential equations by means of the Lie-Bäcklund transformation groups admitted by the given system. This analysis can be effected by extending some of the notions of the classical group analysis of differential equations (Ovsjannikov [2]) to Lie-Bäcklund transformation groups admitted by systems of differential equations. For example, the notion of invariant solutions of differential equations with respect to Lie point transformation groups can be directly extended to the case of Lie-Bäcklund transformation groups.

In order to illustrate these notions, consider the invariance of the time-independent Schrödinger equation for the bound states of the hydrogen atom:

$$-\tfrac{1}{2}\Delta\psi(x)+\left[K+U(x)\right]\psi(x)=0, \tag{13.9}$$

where in appropriate units

$$U(x)=-\frac{1}{\sqrt{(x^1)^2+(x^2)^2+(x^3)^2}},$$

$$\Delta=\left(\frac{\partial}{\partial x^1}\right)^2+\left(\frac{\partial}{\partial x^2}\right)^2+\left(\frac{\partial}{\partial x^3}\right)^2,$$

and $K>0$ is a constant. In this case we have for (13.3)

$$\omega=\tfrac{1}{2}(u_{11}+u_{22}+u_{33})+\left(\frac{1}{r}-K\right)u=0,$$

$$D_i\omega=\tfrac{1}{2}(u_{i11}+u_{i22}+u_{i33})+\left(\frac{1}{r}-K\right)u_i-\frac{x^i}{r^3}u=0, \qquad i=1,2,3,$$

$$\begin{aligned}D_jD_i\omega=\tfrac{1}{2}(u_{ji11}+u_{ji22}+u_{ji33})&-\frac{x^ju_i}{r^3}\\ &+(\frac{1}{r}-K)u_{ij}-\frac{\delta_j^iu}{r^3}\\ &+\frac{3x^jx^iu}{r^5}-\frac{x^iu_j}{r^3}=0, \qquad i,j=1,2,3,\end{aligned} \tag{13.10}$$

$$\cdots\cdots\cdots\cdots\cdots\cdots$$

where

$$r=\sqrt{(x^1)^2+(x^2)^2+(x^2)^3}$$

and

$$u=u^1=\psi,$$

$$u_i=u_i^1, \quad \text{etc.}$$

Now the invariance conditions (13.8) are satisfied by the known O(4) invariance algebra of Fock [1], but here represented as Lie-Bäcklund tangent transformations. In particular, one can verify by direct substitution that the infinitesimal operators with the following coordinates satisfy (13.8):

$$\xi^k_{(i)}=\varepsilon_{ijk}x^j,\quad \eta_{(i)}=0,\cdots,\qquad i,j,k=1,2,3, \tag{13.11}$$

and

$$\left.\begin{aligned}&\xi^k_{(i)}=(2K)^{-1/2}\delta_{ik},\\ &\eta_{(i)}=(2K)^{-1/2}\left\{-\varepsilon_{ijk}\varepsilon_{jlm}x^l u_{mk}+\frac{x^i u}{r}\right\},\\ &\cdots\cdots\cdots\cdots\cdots\cdots\end{aligned}\right\}\qquad i,j,k=1,2,3, \tag{13.12}$$

where δ_{ik} is the Kronecker delta symbol,
$\varepsilon_{ijk}=0$ unless i,j,k are all different,

$$\varepsilon_{ijk}=\begin{cases}+1 \text{ for even permutations of } 1, 2, 3,\\ -1 \text{ for odd permutations of } 1, 2, 3,\end{cases}$$

and the associated ζ_i, $\zeta_{i_1i_2},\cdots$ are defined by (11.13). The first set of three operators corresponds to the set of components of the quantum-mechanical angular-momentum operator, while the second set of three corresponds to the set of components of the quantum-mechanical analogue of the Runge-Lenz vector. The formula (13.11) represents a group of Lie point transformations, i.e. the group of rotations, while (13.12) provides an example of a Lie-Bäcklund invariance group admitted by (13.9).

The connection between the usual form of these operators employed in quantum mechanics and the above form can be established in the following manner. Consider the transformation properties of local manifolds in (x, u)-space under the group G of point transformations (13.1). Suppose that

$$u^\alpha-\phi^\alpha(x)=0$$

transforms under G into the manifold

$$u'^\alpha-\phi'^\alpha(x')=0.$$

Using the transformation law (13.1), the last equation can be written in infinitesimal language in the form

$$u^\alpha-\phi'^\alpha(x)+\left[\eta^\alpha\big(x,u,\underset{1}{u},\cdots\big)-\xi^i\big(x,u,\underset{1}{u},\cdots\big)\frac{\partial\phi'^\alpha(x)}{\partial x^i}\right]a+o(a)=0.$$

This formula is valid for every point (x, u) of the original manifold, so we can replace u^α and $\underset{1}{u},\cdots$ in this expression by $\phi^\alpha(x)$ and its derivatives. This shows that $\phi'^\alpha(x)=\phi^\alpha(x)+o(a)$. Using this fact, and because the function $\phi^\alpha(x)$ is an arbitrary one, we obtain

$$u'^\alpha(x)-u^\alpha(x)=\left[\eta^\alpha\big(x,u(x),\underset{1}{u}(x),\cdots\big)-\xi^i\big(x,u(x),\underset{1}{u}(x),\cdots\big)\frac{\partial u^\alpha(x)}{\partial x^i}\right]a+o(a).$$

We can write this formula in the form

$$u'(x)-u(x)=a\hat{Q}[u(x)]+o(a),$$

where $\hat{Q}=(\hat{Q}^1,\cdots\hat{Q}^m)$, and

$$\hat{Q}^\alpha[u(x)]=\eta^\alpha\left(x,u(x),\frac{\partial u^\beta}{\partial x^j}(x),\cdots\right)-\xi^i\left(x,u(x),\frac{\partial u^\beta(x)}{\partial x^j},\cdots\right)\frac{\partial u^\alpha(x)}{\partial x^i}. \tag{13.13}$$

If in particular the functions ξ^i and η^α are given by the forms

$$\xi^i=\xi^i(x),$$
$$\eta^\alpha=\mu^\alpha(x)+\mu^\alpha_\beta(x)\,u^\beta(x)+\sum_{s=1}^{\infty}\mu^{\alpha,j_1\cdots j_s}_\beta(x)\,u^\beta_{j_1\cdots j_s},$$

then $\hat{Q}$ is an infinite-order linear differential operator of the following form:

$$\hat{Q}^\alpha u(x)=q^\alpha(x)+q^\alpha_\beta(x)u^\beta(x)+\sum_{s=1}^{\infty}q^{\alpha,i_1\cdots i_s}_\beta(x)\frac{\partial^s u^\beta(x)}{\partial x^1\cdots\partial x^s},$$

and is the linear form which appears e.g. in quantum mechanics.

The preceding argument, in particular (13.13), when applied to the Lie-Bäcklund operators comprising the O(4) invariance algebra of Fock [(13.11), (13.12)], yield

$$\hat{Q}_{(i)}=-\varepsilon_{ijk}x^j\frac{\partial}{\partial x^k},\qquad i=1,2,3, \tag{13.14}$$

$$\begin{aligned}\hat{Q}_{(i)}&=(2K)^{-1/2}\left\{-\varepsilon_{ijk}\varepsilon_{jlm}x^l\frac{\partial}{\partial x^m}\frac{\partial}{\partial x^k}+\frac{x^i}{r}\right\}-(2K)^{-1/2}\frac{\partial}{\partial x^i}\\&=(2K)^{-1/2}\left\{\tfrac{1}{2}(\mathbf{L}\times\mathbf{p}-\mathbf{p}\times\mathbf{L})+\frac{\mathbf{x}}{r}\right\}_{(i)},\qquad i=1,2,3.\end{aligned} \tag{13.15}$$

where

$$L_k=\varepsilon_{klm}x^lp_m,\qquad k=1,2,3,$$

and

$$p_k=\frac{1}{i}\frac{\partial}{\partial x^k},\qquad k=1,2,3.$$

§14. Group-Theoretical Nature of Conservation Laws

In classical mechanics it has long been known that the existence of conservation laws for equations of motion is connected with the symmetry properties of the corresponding mechanical systems. In fundamental papers it was shown by Klein [1] and Noether [1] that if a system of differential equations is derivable from a variational principle, there exists a regular procedure for obtaining conservation laws which is based on the investigation of the invariance properties of the variational integral under the action of Lie transformation groups.

In this section we review Noether's classical theorem, which provides sufficient conditions for the existence of conservation laws for the situation in which the variational principle is invariant under the action of Lie transformation groups. Then we consider two levels of generalization of this classical result. The first level of generalization is achieved by focusing on the invariance properties of the extremal values of the variational integrals instead of all their admissible values. This weaker invariance condition leads to the necessary and sufficient conditions for the validity of the conservation laws considered in Noether's work (Ibragimov [3], [4], [5]). This development has been extensively illustrated with examples.

Recently (Ibragimov [6], [7]) a second level of generalization has been realized, which is based on the notion of a weak Lagrangian and the concept of a group of Lie-Bäcklund tangent transformations as developed in Chapter 2. Not only is this result more general in that it incorporates the existence of Lie-Bäcklund tangent transformations, but through the new notion of a weak Lagrangian it is applicable to any system of differential equations, independently of whether the system is or is not derivable from a variational principle.

The following standard definition of a conserved quantity is used in this section. Consider a system of partial differential equations

$$(\omega) \qquad \omega_\nu\left(x, u, \underset{1}{u}, \cdots, \underset{q}{u}\right)=0, \qquad \nu=1,\cdots,N,$$

where

$$x=(x^1, x^2,\cdots,x^n),$$

$$u=(u^1, u^2,\cdots,u^m),$$

$$\underset{1}{u}=\{u_i^\alpha \,|\, i=1,\cdots,n,\ \alpha=1,\cdots,m\},$$

$$\underset{2}{u}=\{u_{ij}^\alpha \,|\, i,j=1,\cdots,n,\ \alpha=1,\cdots,m\},$$

$$\cdots\cdots\cdots\cdots\cdots\cdots\cdots\cdots$$

In the space of all sequences $(x, u, \underset{1}{u}, \cdots)$, let Ω denote the manifold defined by the equations

$$(\Omega) \qquad \left\{\begin{array}{r} \omega_\nu = 0, \\ D_i\omega_\nu = 0, \\ D_iD_j\omega_\nu = 0, \\ \cdots\cdots \end{array}\right\} \qquad \nu = 1, \cdots, N,$$

where

$$D_i = \frac{\partial}{\partial x^i} + u_i^\alpha \frac{\partial}{\partial u^\alpha} + u_{ij}^\alpha \frac{\partial}{\partial u_j^\alpha} + \cdots. \tag{14.1}$$

DEFINITION 14.1. A vector $A = (A^1, \cdots, A^n)$ is conserved by the system (ω) if

$$D_i A^i|_\Omega = 0. \tag{14.2}$$

The vector $A = (A^1, \cdots, A^n)$ will be called a *c*-vector or, if $n = 1$, a *c*-quantity.

Such conserved quantities were first extensively studied by Felix Klein (1918) in connection with the theory of general relativity and by Emmy Noether (1918) in relation to the Euler equations which arise in the solution of variational problems.

§14.1 Lagrangian Structure. Let us consider here functionals of the form

$$l(u) = \int \mathfrak{L}\left(x, u, \underset{1}{u}\right) dx, \tag{14.3}$$

where the Lagrangian $\mathfrak{L}$ depends only on x, u and $\underset{1}{u}$. The extrema of $l(u)$ will then satisfy the Euler equations

$$\frac{\delta \mathfrak{L}}{\delta u^\alpha} = \frac{\partial \mathfrak{L}}{\partial u^\alpha} - D_i \frac{\partial \mathfrak{L}}{\partial u_i^\alpha} = 0, \qquad \alpha = 1, \cdots, m. \tag{14.4}$$

DEFINITION 14.2. The variational integral (14.3) is called *nondegenerate* if all equations of the system (14.4) are second-order ones.

The Euler equations (14.4) are a system of quasilinear equations of the form

$$a^{ij}(x, u, \partial u) u_{ij} + b(x, u, \partial u) = 0,$$

where a^{ij}, $i, j = 1, \cdots, n$, are $m \times m$ matrices and $b = (b^1, \cdots, b^m)$. The nondegenerate conditions means that

$$\operatorname{rank}\|a^{11} \cdots a^{ij} \cdots a^{nn}\| = m.$$

Now suppose a one-parameter group of transformations G acts on (x, u) by means of the equations:

$$\begin{aligned} x'^i &= f^i(x, u, a), \\ u'^\alpha &= \phi^\alpha(x, u, a). \end{aligned} \tag{14.5}$$

The group G may be extended so as to encompass the transformation of the

u_i^α. Therefore, given a functional of the form (14.3), we may define

$$l'(u')=\int \mathfrak{L}\left(x', u', \underset{1}{u'}\right)dx'. \tag{14.6}$$

DEFINITION 14.3. The functional l is *invariant* under the action of G if $l(u)=l'(u')$ for every function u and parameter a.

Noether's theorem then establishes a sufficient condition for the existence of a conserved vector for the equations (14.4).

THEOREM 14.1 (Noether [1]). *If l is invariant under G, then there exists a conservation law for the corresponding Euler equations* (14.4). *Indeed, the conserved vector is given by*

$$A^i=\left(\eta^\alpha-u_j^\alpha\xi^j\right)\frac{\partial\mathfrak{L}}{\partial u_i^\alpha}+\mathfrak{L}\xi^i. \tag{14.7}$$

The invariance of l under G is, however, a very strong requirement. The following theorem gives the necessary and sufficient condition for the existence of such a conserved quantity for the Euler equations (14.4).

THEOREM 14.2 (Ibragimov [3]). *Suppose $\delta\mathfrak{L}/\delta u^\alpha=0$ is invariant under G. Then the vector A, with components given by* (14.7), *satisfies the conservation equation* (14.2) *if and only if the extremal values of l are invariant under G.*

Thus for systems of the type (14.4) which are invariant under some group, the criteria for the existence of a conserved vector of the form (14.7) are known.

Noether's theorem (Theorem 14.1) is a particular case of Theorem 14.2. Indeed, if the integral (14.3) is invariant with respect to a group G, the Euler equations (14.4) admit the group G. In Theorem 14.1 the linear independence of the c-vectors which are defined in (14.7), is not guaranteed. Nevertheless, all conservation laws which one obtains from Theorem 14.1 may be obtained from Theorem 14.2. Therefore, the latter gives no fewer independent c-vectors than the former. The two theorems are identical in the case that (14.4) is nondegenerate, and this fact is expressed in the following theorem:

THEOREM 14.3 (Ibragimov [3]). *Let the variational integral* (14.3) *be nondegenerate and the Euler equations* (14.4) *admit the group G of transformations* (14.5). *Then the conservation equation* (14.2) *is valid if and only if the integral* (14.3) *is invariant with respect to the group G.*

Finally, we close this subsection with a brief remark on another known simple generalization of Theorem 14.1. This follows from the so-called divergence transformations. One says that the transformations (14.5) are divergence transformations if for the associated tangent vector the equation

$$\xi^i\frac{\partial\mathfrak{L}}{\partial x^i}+\eta^\alpha\frac{\partial\mathfrak{L}}{\partial u^\alpha}+\left(D_i\eta^\alpha-u_j^\alpha D_i\xi^j\right)\frac{\partial\mathfrak{L}}{\partial u_i^\alpha}+\mathfrak{L}D_i\xi^i=D_iB_i \tag{14.8}$$

is valid for some vector $B(x, u, \partial u, \cdots)$. If the right side of (14.8) does not vanish, the variational integral (14.3) will not be invariant under transformations (14.5), so the conservation equation (14.2) is not valid. However, for the vector

$$C = A - B, \tag{14.9}$$

the conservation equation (14.2) is valid where the vector A is defined by the formula (14.7) and B by (14.8).

§14.2. Examples. Now let us turn to several illustrations of the general conservation theorems discussed in §14.1. The examples presented in this section have been taken from Ibragimov [5]. The equations for the motion of a free particle in De Sitter space and for unsteady transonic gas motion illustrate nondegenerate variational integrals, while the Dirac equation illustrates the case of a degenerate variational integral. It is also shown that in the case of the conservation laws of fluid mechanics Noether's theorem is useful even though the equations under discussion are not of the form (14.4). We conclude this subsection with comments on higher-order variational problems.

Example. Motion of a Free particle in De Sitter space. A free particle in 4-dimensional space-time V_4 moves along a geodesic. Let $x^\alpha = x^\alpha\ (\sigma)$, $\alpha = 1, \cdots, 4$, be a curve in V_4 with the fundamental form

$$ds^2 = g_{\alpha\beta}\,(x)\,dx^\alpha\,dx^\beta,$$

and

$$\dot{x}^\alpha = \frac{dx^\alpha}{d\sigma}, \qquad \alpha = 1, \cdots, 4.$$

The equations of the particle motion are derived from the variational principle with the Lagrangian function

$$\mathfrak{L} = -mc\sqrt{g_{\alpha\beta}(x)\,\dot{x}^\alpha\,\dot{x}^\beta}\,, \tag{14.10}$$

where c is the velocity of light and m is the mass of the particle. The parameter σ is the independent variable, and the coordinates of the 4-vector $x = (x^1, \cdots, x^4)$ are dependent ones.

The variational integral (14.3) with the Lagrangian function (14.10) is invariant with respect to the group of motions of V_4. The tangent vector for this group is of the form

$$x = \eta^\alpha \frac{\partial}{\partial x^\alpha}, \tag{14.11}$$

where $\eta = (\eta^1, \eta^2, \eta^3, \eta^4)$ is the solution of the Killing equations

$$\eta_{\alpha,\,\beta} + \eta_{\beta,\,\alpha} = 0, \qquad \alpha, \beta = 1, \cdots, 4.$$

The comma in this formula denotes covariant differentiation. Let us choose the length s of a curve as the parameter σ. Then we shall obtain the c-quantity in the form

$$A = mcg_{\alpha\beta}\,(x)\,\dot{x}^\alpha\eta^\beta \tag{14.12}$$

for every solution η of the Killing equations satisfying the formula (14.2).

We shall consider a space-time V_4 of constant curvature known as a De Sitter space. The group of motions in a De Sitter space (the De Sitter group) is a 10-parameter group, so 10 independent conservation laws exist for free motion of a particle in a De Sitter space, as is the case in relativistic mechanics. Below we shall write out all the c-quantities, with the index K indicating these conservation laws corresponding to the motion in a De Sitter space with constant curvature K. In particular, c-quantities in relativistic mechanics will have the index 0.

The fundamental form for De Sitter space can be written in the canonical form

$$ds^2 = H^{-2}(c^2dt^2 - dx^2 - dy^2 - dz^2), \tag{14.13}$$

where

$$H = 1 + \frac{K}{4}(c^2t^2 - x^2 - y^2 - z^2).$$

Below, the notation $x = x^1, y = x^2, z = x^3, t = x^4$ will be used. The Latin indexes i, j, k will run from 1 to 3, and Greek indexes α, β from 1 to 4. Boldface will be used for spatial vectors: $\mathbf{x} = (x^1, x^2, x^3)$, $\mathbf{x}\cdot\mathbf{v} = \Sigma_1^3 x^i v^i$, and $\mathbf{x}\times\mathbf{v}$ is the vector product: $(\mathbf{x}\times\mathbf{v})^i = \varepsilon_{ijk}x^j v^k$, where ε_{ijk} is the usual permutation symbol.

The components $\dot{x}^\alpha = dx^\alpha/ds$ of the 4-velocity in De Sitter space and the components $v^i = dx^i/dt$ of the physical velocity $\mathbf{v}$ are connected by

$$\dot{x}^i = \frac{Hv^i}{\sqrt{1-\beta^2}}, \quad \dot{x}^4 = \frac{H}{c\sqrt{1-\beta^2}}, \qquad \text{where } \beta^2 = \frac{|\mathbf{v}|^2}{c^2}. \tag{14.14}$$

Note that in this subsection β is employed both as an index and in the above sense, but it should be clear from the context which meaning is intended.

Now let's consider the different subgroups of the De Sitter group together with their corresponding c-quantities.

1. *Conservation of momentum.* In classical and relativistic mechanics the conservation of momentum arises because of the invariance of a mechanical system under arbitrary translations of the axes. In De Sitter space, instead of the translation group, we have the 3-parameter group, generated by tangent vectors of the form (14.11) with

$$\eta_i^\alpha = (H-2)\delta^{i\alpha} + \frac{K}{2}x^i x^\alpha. \tag{14.15}$$

Substituting (14.15) into the formula (14.12) and using the formulae (14.13) and (14.14), we obtain the c-quantities

$$A_i = \frac{m}{H\sqrt{1-\beta^2}}\left\{(2-H)v^i + \frac{K}{2}(c^2t^2 - \mathbf{x}\cdot\mathbf{v})x^i\right\}.$$

So, the momentum of a free particle in a De Sitter space is defined by the formula

$$p_K = \frac{m}{\sqrt{1-\beta^2}} \left\{ \left(\frac{2}{H} - 1 \right) \mathbf{v} + \frac{K}{2H} (c^2 t^2 - \mathbf{x} \cdot \mathbf{v}) \mathbf{x} \right\}. \tag{14.16}$$

If $K=0$, this formula gives the well-known formula of the momentum in relativistic mechanics:

$$\mathbf{p}_0 = \frac{m\mathbf{v}}{\sqrt{1-\beta^2}}.$$

2. *Conservation of energy.* The conservation of energy is the result of the invariance of the variational integral under the one-parameter subgroup of the De Sitter group generated by the tangent vector (14.11) with

$$\eta^i = \frac{K}{2} x^4 x^i, \qquad \eta^4 = \frac{1}{c^2} \left(H + \frac{K}{2} |\mathbf{x}|^2 \right). \tag{14.17}$$

The c-quantity (14.12) in this case is equal to

$$A = \frac{m}{\sqrt{1-\beta^2}} \left\{ 1 + \frac{K}{2H} (\mathbf{x} - t\mathbf{v}) \cdot \mathbf{x} \right\}.$$

This quantity multiplied by $-c^2$ is the energy of a free particle in a De Sitter space:

$$E_K = \frac{mc^2}{\sqrt{1-\beta^2}} \left\{ 1 + \frac{K}{2H} (\mathbf{x} - t\mathbf{v}) \cdot \mathbf{x} \right\}. \tag{14.18}$$

For $K=0$ we have the formula for the relativistic energy,

$$E_0 = \frac{mc^2}{\sqrt{1-\beta^2}}.$$

3. *Conservation of angular momentum.* The De Sitter group contains the subgroup of rotations of the axes and the subgroup of the Lorentz transformations. The conservation of angular momentum corresponds to the invariance of the variational integral with respect to rotations with tangent vectors of the type (14.11), where

$$\eta_i^j = e_{ijk} x^k, \qquad \eta_i^4 = 0. \tag{14.19}$$

The expression for angular momentum according to (14.12), (14.16) and the identity $\mathbf{x} \times \mathbf{x} = 0$ can be written in the form

$$M_K = \frac{1}{2-H} (\mathbf{x} \times \mathbf{p}_K). \tag{14.20}$$

In particular, $M_0 = \mathbf{x} \times \mathbf{p}_0$.

4. *Center-of-mass theorem.* The center-of-mass theorem for the N-body problem of relativistic mechanics is associated with the subgroup of Lorentz transfor-

mations. If $N=1$, this theorem is equivalent to the statement that the conservation equation $dQ_0/dt=0$ for the vector

$$Q_0=\frac{m}{\sqrt{1-\beta^2}}(\mathbf{x}-t\mathbf{v}),$$

is valid. In De Sitter space the analogous vector-valued c-quantity is given by the formula

$$Q_K=\frac{mH^{-4}}{\sqrt{1-\beta^2}}(\mathbf{x}-\mathbf{v}). \tag{14.21}$$

Example. Unsteady transonic gas motion. For the equation for unsteady transonic gas motion,

$$-\Phi_x\Phi_{xx}-2\Phi_{xt}+\Phi_{yy}=0, \tag{14.22}$$

the Lagrangian function can be taken in the form

$$\mathfrak{L}=-\tfrac{1}{6}\Phi_x^3-\Phi_x\Phi_t+\tfrac{1}{2}\Phi_y^2. \tag{14.23}$$

Equation (14.22) admits the infinite group G_∞ characterized by the tangent vectors

$$X=\xi^0\frac{\partial}{\partial t}+\xi^1\frac{\partial}{\partial x}+\xi^2\frac{\partial}{\partial y}+\eta\frac{\partial}{\partial\Phi}, \tag{14.24}$$

which depend on five arbitrary functions of time $f(t)$, $g(t)$, $h(t)$, $\sigma(t)$, $\tau(t)$ and have the coordinates

$$\begin{aligned}
&\xi^0=3f(t),\\
&\xi^1=f'(t)x+f''(t)y^2+g'(t)y+h(t), \qquad f'\equiv\frac{df}{dt},\dots\\
&\xi^2=2f'(t)y+g(t),\\
&\eta=-f'(t)\Phi+f''(t)x^2+2f'''(t)xy^2+\tfrac{1}{3}f^{(4)}(t)y^4+2g''(t)xy\\
&\qquad+\tfrac{2}{3}g'''(t)y^3+2h'(t)x+2h''(t)y^2+\sigma(t)y+\tau(t).
\end{aligned} \tag{14.25}$$

The group G_∞ is an example of the divergence transformations discussed in §14.1. In particular, let us consider the subgroup of the group G_∞ generated by the tangent vector (14.24), where

$$\xi^0=0,\quad \xi^1=0,\quad \xi^2=0,\quad \eta=\sigma(t)y. \tag{14.26}$$

The conservation equations will be written in the form

$$D_tC^1+D_xC^2+D_yC^3=0.$$

In accordance with (14.7),

$$A^1=-\sigma y\Phi_x,\quad A^2=-\sigma y\left(\Phi_t+\tfrac{1}{2}\Phi_x^2\right),\quad A^3=\sigma y\Phi_y.$$

The left side of (14.8) for (14.23), (14.26) is equal to

$$-\sigma' y\Phi_x+\sigma\Phi_y=D_x(-\sigma' y\Phi)+D_y(\sigma\Phi),$$

so in the right side of (14.8) one can take the vector **B** with the components

$$B^1=0,\quad B^2=-\sigma' y\Phi,\quad B^3=\sigma\Phi.$$

Substitution of these vectors **A** and **B** in (14.9) yields the conservation equation

$$D_t(-\sigma y\Phi_x)+D_x(-\sigma y\Phi_t-\tfrac{1}{2}\sigma y\Phi_x^2+\sigma' y\Phi)+D_y(\sigma y\Phi_y+\sigma\Phi)=0. \tag{14.27}$$

The analogous calculations, carried out for the general tangent vector (14.25) of the group G_∞, yield a c-vector depending on five arbitrary functions of time. The coordinates of this vector are

$$\begin{aligned}
C^1=&-\tfrac{1}{2}f\Phi_x^3+(f'x+f''y^2+g'y+h)\Phi_x^2+\tfrac{3}{2}f\Phi_y^2+(2f'y+g)\Phi_x\Phi_y\\
&+\big(f'\Phi-f''x^2-2f'''xy^2-\tfrac{1}{3}f^{(4)}y^4-2g''xy\\
&-\tfrac{2}{3}g'''y^3-2h'x-2h''y^2-\sigma y-\tau\big)\Phi_x\\
&+2\,(f''x+f'''y^2+g''y+h')\Phi,\\
C^2=&\tfrac{1}{3}\,(f'x+f''y^2+g'y+h)\big(\Phi_x^3+\tfrac{3}{2}\Phi_y^2\big)\\
&+\big(3f\Phi_t+2f'y\Phi_y+g\Phi_y+f'\Phi-f''x^2-2f'''xy^2\\
&-\tfrac{1}{3}f^{(4)}y^4-2g''xy-\tfrac{2}{3}g'''y^3-2h'x-2h''y^2-\sigma y-\tau\big)\big(\Phi_t+\tfrac{1}{2}\Phi_x^2\big)\\
&-\tfrac{1}{2}f''\Phi^2+\big(f'''x^2+2f^{(4)}xy^2+\tfrac{1}{3}f^{(5)}y^4+2g'''xy+\tfrac{2}{3}g^{(4)}y^3\\
&+2h''x+2h'''y^2+\sigma' y+\tau'\big)\Phi,\\
C^3=&-\big(f'y+\tfrac{1}{2}g\big)\big(\tfrac{1}{3}\Phi_x^3+\Phi_y^2+2\Phi_x\Phi_t\big)\\
&-(f'x+f''y^2+g'y+h)\Phi_x\Phi_y-3f\Phi_y\Phi_t\\
&+\big(f''x^2+2f'''xy^2+\tfrac{1}{3}f^{(4)}y^4-f'\Phi+2g''xy\\
&+\tfrac{2}{3}g'''y^3+2h'x+2h''y^2+\sigma y+\tau\big)\Phi_y\\
&-2\,\big(2f'''xy+\tfrac{2}{3}f^{(4)}y^3+g''x+g'''y^2+2h''y+\tfrac{1}{2}\sigma\big)\Phi.
\end{aligned}$$

Example. Dirac equations.. The Dirac equations

$$\gamma^k\frac{\partial}{\partial x^k}+m\psi=0 \tag{14.28}$$

together with the conjugate equations

$$\frac{\partial\tilde\psi}{\partial x^k}\gamma^k-m\tilde\psi=0 \tag{14.29}$$

are derivable from the variational principle with the Lagrangian function

(14.30) $$\mathfrak{L}=\frac{1}{2.}\left\{\tilde{\psi}\left(\gamma^k\frac{\partial\psi}{\partial x^k}+m\psi\right)-\left(\frac{\partial\tilde{\psi}}{\partial x^k}\gamma^k-m\tilde{\psi}\right)\psi\right\}.$$

Here ψ is a complex 4-dimensional column vector, γ^k are the Dirac matrices

$$\gamma^1=\begin{Vmatrix}0&0&0&-i\\0&0&-i&0\\0&i&0&0\\i&0&0&0\end{Vmatrix},\qquad \gamma^2=\begin{Vmatrix}0&0&0&-1\\0&0&1&0\\0&1&0&0\\-1&0&0&0\end{Vmatrix},$$

$$\gamma^3=\begin{Vmatrix}0&0&-i&0\\0&0&0&i\\i&0&0&0\\0&-i&0&0\end{Vmatrix},\qquad \gamma^4=\begin{Vmatrix}1&0&0&0\\0&1&0&0\\0&0&-1&0\\0&0&0&-1\end{Vmatrix},$$

and $\tilde{\psi}$ is a row vector defined by the formula

(14.31) $$\tilde{\psi}=\bar{\psi}^T\gamma^4,$$

where ψ^T is a row vector, complex conjugate with the vector ψ.

The Dirac equations give us an example of the degenerate variational problem. To illustrate the difference between Theorem 14.1 and Theorem 14.2, let us consider the following simple group of transformations admissible by the Dirac equations. Let G_∞ be infinite group generated by the tangent vectors

(14.32) $$x=\eta^\alpha(x)\frac{\partial}{\partial\psi^\alpha}+\tilde{\eta}^\alpha(x)\frac{\partial}{\partial\bar{\psi}^\alpha},$$

where the vectors $\eta(x)$, $\tilde{\eta}(x)$ are connected by means of (14.31) and satisfy the Dirac equations (14.28) and (14.29). The extremals of the variational integral (14.3) with the Lagrangian function (14.30) are invariant with respect to the group G_∞. It is easy to verify that the variational integral is not invariant with respect to this group. According to Theorem 14.2, substitution of the expressions (14.30), (14.32) into (14.7) gives the c-vector A_∞ with the components

$$A^k_\infty=\tilde{\psi}\gamma^k X(x)-\tilde{X}(x)\gamma^k\psi,\qquad k=1,\cdots,4.$$

As is well known, the Dirac equations with zero rest mass admit the 15-parameter conformal group. This group is generated by the tangent vectors

(14.33) $$X=\xi^i\frac{\partial}{\partial x^i}+\eta^\alpha\frac{\partial}{\partial\psi^\alpha}+\tilde{\eta}^\alpha\frac{\partial}{\partial\bar{\psi}^\alpha},$$

where ξ is the solution of the generalized Killing equations,

$$\frac{\partial\xi^j}{\partial x^k}+\frac{\partial\xi^k}{\partial x^j}=\mu(x)\delta^{jk},\qquad j,k=1,\cdots,4,$$

and η is given by

$$\eta = S\psi, \qquad \tilde{\eta} = \tilde{\psi}\tilde{S},$$

with

$$S = \frac{1}{8} \sum_{j,k=1}^{4} \frac{\partial \xi^j}{\partial x^k}(\gamma^j\gamma^k - \gamma^k\gamma^j - 3\delta^{jk}), \qquad \tilde{S} = \gamma^4 \bar{S}^T \gamma^4.$$

The invariance of the Dirac equations with respect to the conformal group yields the following 15 independent c-vectors: A_k, A_{kp} $(k<p)$, B_k $(k, p = 1,\cdots,4)$, A_0 with the components

$$A_k^j = \frac{1}{2}\left\{ \frac{\partial \tilde{\psi}}{\partial x^k}\gamma^j\psi - \tilde{\psi}\gamma^j\frac{\partial \psi}{\partial x^k} + \delta_k^j\left(\tilde{\psi}\gamma^p\frac{\partial \psi}{\partial x^p} - \frac{\partial \tilde{\psi}}{\partial x^p}\gamma^p\psi\right)\right\}, \tag{14.34}$$

$$A_{kp}^j = \tfrac{1}{4}\{\psi\,(\gamma^j\gamma^k\gamma^p + \gamma^k\gamma^p\gamma^j)\psi\} + x^p A_k^j - x^k A_p^j, \tag{14.35}$$

$$B_k^j = 2x^p A_{pk}^j + |x|^2 A_k^j, \qquad |x|^2 = \sum_{j=1}^{4}(x^j)^2, \tag{14.36}$$

$$A_0^j = x^k A_k^j. \tag{14.37}$$

The Dirac equations with arbitrary rest mass m also admit a 4-parameter group generated by the following one-parameter groups of transformations:

$$\begin{aligned} &\psi' = a\psi, && \psi' = \psi e^{-ia}, \\ &\psi' = \psi\cosh a + \gamma^4\gamma^2\tilde{\psi}^T\sinh a, && \psi' = \psi\cosh a + i\gamma^4\gamma^2\tilde{\psi}^T\sinh a. \end{aligned} \tag{14.38}$$

In this and later formulae only the transformations of the vector ψ are written. The transformations of the vector $\tilde{\psi}$ are defined by the formula (14.31).

The Dirac equations with zero rest mass admit the 4-parameter Pauli group as well. This group is generated by the following one-parameter groups of transformations:

$$\begin{aligned} &\psi' = \psi e^{ia\gamma^5}, && \psi' = \psi e^{-a\gamma^5}, \\ &\psi' = \psi\cos a + \gamma^3\gamma^1\tilde{\psi}^T\sin a, && \psi' = \psi\cos a + i\gamma^3\gamma^1\tilde{\psi}^T\sin a, \end{aligned}$$

where $\gamma^5 = \gamma^1\gamma^2\gamma^3\gamma^4$.

The transformations mentioned above, together with the group G_∞, constitute the main group of the Dirac equations (i.e., the most general continuous local group of Lie transformations admissible by the Dirac equations (14.28), (14.29) (Ibragimov [5])). The extremals of the variational integral are all invariant with respect to this group, with the exception of the first transformation in (14.38). It should be noted that not all the c-vectors obtained with this group are independent. Namely, the c-vectors associated with the last three Pauli transformations appear to be identically zero, while other transformations yield four independent

c-vectors C_k, $k=1,\cdots 4$, with the components

$$C_1^j=-i\tilde{\psi}\gamma^j\psi,\qquad C_2^j=\tfrac{1}{2}(\tilde{\psi}\gamma^j\gamma^4\gamma^2\tilde{\psi}^T-\psi^T\gamma^4\gamma^2\gamma^j\psi),$$

$$C_3^j=\frac{i}{2}(\tilde{\psi}\gamma^j\gamma^4\gamma^2\tilde{\psi}^T+\psi^T\gamma^4\gamma^2\gamma^j\psi),\qquad C_4^j=i\tilde{\psi}\gamma^j\gamma^5\psi.$$

If $m\neq 0$, the full list of c-vectors consists of A_∞, A_k, A_{kp} $(k,p=1,\cdots,4)$, C_1, C_2, C_3.

Example. Fluid mechanics.

1. *The ideal polytropic gas flow.* Let us consider the equations governing the motion of an ideal polytropic gas,

$$\begin{aligned}&\mathbf{v}_t+(\mathbf{v}\cdot\nabla)\mathbf{v}+\frac{1}{\rho}\nabla p=0,\\ &\rho_t+\mathbf{v}\cdot\nabla\rho+\rho\operatorname{div}\mathbf{v}=0,\\ &p_t+\mathbf{v}\cdot\nabla p+\gamma p\operatorname{div}\mathbf{v}=0,\qquad \gamma\equiv c_p/c_v=\text{const},\end{aligned}\tag{14.39}$$

where the velocity vector $\mathbf{v}$ (with components $v^1,\cdots,v^n$), pressure p, and density ρ are functions of the variables t and cartesian coordinate vector $\mathbf{x}$ with components $x^1,\cdots,x^n$. The symbols ∇ and div represent the gradient operator and divergence with respect to $\mathbf{x}$.

The Lie transformation group admitted by (14.39) is known to be larger than that for the general perfect-gas motion equations (Ovsjannikov [2]). It is also known that when

$$\gamma=\frac{n+2}{n},\tag{14.40}$$

there is an enlargement of the admissible group for (14.39). Here $n=1$, 2, 3 correspond to the one-dimensional, planar, and three-dimensional flows, respectively. We shall sketch the way in which the group properties of the equations of motion and the general conservation theorems were applied to obtain conservation laws for an ideal gas and a perfect incompressible fluid. First, let us consider the particular case of an isentropic potential flow of an ideal polytropic gas. This also serves as a model for all other cases. In this case we can consider the equation for the potential Φ $(t,\mathbf{x})$,

$$\Phi_{tt}+2\nabla\Phi\cdot\nabla\phi_t+\nabla\Phi\cdot(\nabla\Phi\cdot\nabla)\nabla\Phi+(\gamma-1)(\Phi_t+\tfrac{1}{2}|\nabla\Phi|^2)\Delta\Phi=0,\tag{14.41}$$

instead of (14.39). Equation (14.41) is derivable by the application of the variational procedure to the variational integral (14.3) with the Lagrangian function

$$\mathfrak{L}=\left(\Phi_t+\tfrac{1}{2}|\nabla\Phi|^2\right)^{\gamma/(\gamma-1)}.$$

Now a calculation of the Lie transformation group admitted by (14.41) plus an application of the Noether Theorem 14.1 yield conservation laws for (14.41). Then we rewrite these conservation laws in the variables $\mathbf{v}$, p, ρ using the definition of the potential Φ; the Lagrange-Cauchy integral, which can be taken in the form

$$\Phi_t + \tfrac{1}{2}|\nabla\Phi|^2 + \frac{\gamma}{\gamma-1}\rho^{\gamma-1} = 0;$$

and the relation between pressure p and density ρ in the isentropic flow of the ideal polytropic gas.

Here the integral form of the conservation equations is used, which is equivalent to the differential form (14.2). The following symbols are used:

$\Omega(t)$—arbitrary n-dimensional volume, moving with fluid,
$S(t)$—boundary of the volume $\Omega(t)$,
$\boldsymbol{\nu}$—unit (outer) vector normal to the surface $S(t)$.

The invariance of the equation (14.41) with respect to the one-parameter group of transformations $\Phi' = \Phi + a$ and the Galilei group (translations in the time, translations of the axes, 3-parameter group of rotations of the axes and 3-parameter group of velocity transformations) yield the following classical conservation equations:

Conservation of mass,

$$\frac{d}{dt}\int_{\Omega(t)} \rho\, d\omega = 0.$$

Conservation of energy,

$$\frac{d}{dt}\int_{\Omega(t)} \left(\tfrac{1}{2}\rho|\mathbf{v}|^2 + \frac{p}{\gamma-1}\right) d\omega = -\int_{S(t)} p\mathbf{v}\cdot\boldsymbol{\nu}\, dS.$$

Conservation of momentum,

$$\frac{d}{dt}\int_{\Omega(t)} \rho\mathbf{v}\, d\omega = -\int_{S(t)} p\boldsymbol{\nu}\, dS.$$

Conservation of angular momentum,

$$\frac{d}{dt}\int_{\Omega(t)} \rho\,(\mathbf{x}\times\mathbf{v})\, d\omega = -\int_{S(t)} p\,(\mathbf{x}\times\boldsymbol{\nu})\, dS.$$

Center-of-mass theorem,

$$\frac{d}{dt}\int_{\Omega(t)} \rho\,(t\mathbf{v}-\mathbf{x})\, d\omega = -\int_{S(t)} tp\boldsymbol{\nu}\, dS.$$

The following additional conservation equations are valid only for an ideal polytropic gas satisfying the condition (14.40):

$$\frac{d}{dt}\int_{\Omega(t)}\{t(\rho|\mathbf{v}|^2+np)-\rho\mathbf{x}\cdot\mathbf{v}\}\,d\omega=-\int_{S(t)}p\,(2t\mathbf{v}-\mathbf{x})\cdot\boldsymbol{\nu}\,dS, \tag{14.42}$$

$$\frac{d}{dt}\int_{\Omega(t)}\{t^2(\rho|\mathbf{v}|^2+np)-\rho\mathbf{x}\cdot(2t\mathbf{v}-\mathbf{x})\}\,d\omega=-\int_{S(t)}2tp\,(t\mathbf{v}-\mathbf{x})\cdot\boldsymbol{\nu}\,dS. \tag{14.43}$$

The conservation equations (14.42), (14.43) arise from the invariance of (14.41) with respect to two one-parameter groups generated by the tangent vectors of the form

$$X=\xi^0\frac{\partial}{\partial t}+\xi^i\frac{\partial}{\partial x^i}+\eta\frac{\partial}{\partial\Phi} \tag{14.44}$$

with

$$\xi^0=2t,\quad \xi^i=0\ (i=1,2,3),\quad \eta=0$$

and

$$\xi^0=t^2,\quad \xi^i=tx^i\ (i=1,2,3),\quad \eta=\tfrac{1}{2}|\mathbf{x}|^2,$$

respectively, if the condition (14.40) is satisfied.

More generally for the case of potential flow of an ideal polytropic gas with the arbitrary γ the following conservation equation is valid:

$$\begin{aligned}\frac{d}{dt}\int_{\Omega(t)}\bigg\{&\frac{2\gamma+n(\gamma-1)}{\gamma+1}\,t\left(\tfrac{1}{2}\rho|\mathbf{v}|^2+\frac{p}{\gamma-1}\right)\\ &-\rho\mathbf{x}\cdot\mathbf{v}-\frac{n(\gamma-1)-2}{\gamma+1}\rho\Phi\bigg\}\,d\omega\\ &=-\int_{S(t)}p\left(\frac{2\gamma+n(\gamma-1)}{\gamma+1}\,t\mathbf{v}-\mathbf{x}\right)\cdot\boldsymbol{\nu}\,dS.\end{aligned} \tag{14.45}$$

Under the condition (14.41), equation (14.42) turns into equation (14.43).

For three-dimensional flow the condition (14.40) ($\gamma=\frac{5}{3}$) is satisfied for a monatomic gas. So for a monatomic gas the additional conservation equations (14.42) and (14.43) are valid.

2. *Perfect incompressible fluid flow.* For the equations of motion of a perfect incompressible fluid,

$$\mathbf{v}_t+(\mathbf{v}\cdot\nabla)\mathbf{v}+\nabla p=0,\qquad \operatorname{div}\mathbf{v}=0, \tag{14.46}$$

some additional conservation laws hold. One of these laws is valid for potential flow of the fluid and has the form

$$\frac{d}{dt}\int_{\Omega(t)}\left(\frac{n+2}{2}\,t|\mathbf{v}|^2-\mathbf{x}\cdot\mathbf{v}-n\Phi\right)d\omega=-\int_{S(t)}p\,[(n+2)t\mathbf{v}-\mathbf{x}]\cdot\boldsymbol{\nu}\,dS.$$

Formally, this conservation equation may be obtained from (14.45) assuming $\gamma\to\infty$.

Now, for any vector $\mathbf{v}$ which satisfies the condition $\operatorname{div}\mathbf{v}=0$, the equation

$$\int_\Omega \mathbf{v}\,d\omega=\int_S(\mathbf{v}\cdot\boldsymbol{\nu})\mathbf{x}\,dS$$

is valid. As a result we have the following generalized form of the conservation law of momentum for an incompressible fluid:

$$\frac{d}{dt}\int_{\Omega(t)}(\mathbf{f}\cdot\mathbf{v})\,d\omega=-\int_{S(t)}\left\{p\mathbf{f}\cdot\boldsymbol{\nu}-\left(\frac{d\mathbf{f}}{dt}\cdot\mathbf{x}\right)(\mathbf{v}\cdot\boldsymbol{\nu})\right\}dS, \tag{14.47}$$

where $\mathbf{f}(t)=(f^1(t),\cdots,f^n(t))$ is an arbitrary smooth vector function of time. From the group-theoretical point of view the conservation equation (14.47) is the consequence of the invariance of equations (14.46) with respect to the infinite group of transformations, generated by the tangent vector

$$X=\xi^0\frac{\partial}{\partial t}+\xi^i\frac{\partial}{\partial x^i}+\eta^\alpha\frac{\partial}{\partial v^\alpha}+\tilde{\eta}\frac{\partial}{\partial \rho}, \tag{14.48}$$

where

$$\xi^0=0,\qquad \xi^i=f^i(t),\qquad i=1,\cdots,n,$$
$$\eta^\alpha=\frac{df^\alpha}{dt}(t),\qquad \alpha=1,\cdots,n,$$
$$\tilde{\eta}=-\frac{d^2\mathbf{f}(t)}{dt^2}\cdot\mathbf{x}.$$

3. *Shallow-water flow.* It is known that in the case $n=2$ equations (14.39) can be interpreted as the shallow-water-theory equations, if the condition (14.40) is valid. Namely, if we put

$$p=\tfrac{1}{2}\rho^2,\qquad \rho=gh,$$

where g is the gravitation constant and h is the depth of the fluid, then we obtain the shallow-water-theory equations

$$\mathbf{v}_t+(\mathbf{v}\cdot\nabla)\mathbf{v}+g\nabla h=0,$$
$$h_t+\mathbf{v}\cdot\nabla h+h\operatorname{div}\mathbf{v}=0,$$

and because the equations (14.42) (14.43) apply, the conservation laws

$$\frac{d}{dt}\int_{\Omega(t)}h\left\{t\left(|\mathbf{v}|^2+gh\right)-\mathbf{x}\cdot\mathbf{v}\right\}d\omega=-\int_{S(t)}\tfrac{1}{2}gh^2(2t\mathbf{v}-\mathbf{x})\cdot\boldsymbol{\nu}\,dS \tag{14.49}$$

and

$$\frac{d}{dt}\int_{\Omega(t)}h\left\{t^2(|\mathbf{v}|^2+gh)-\mathbf{x}\cdot(2t\mathbf{v}-\mathbf{x})\right\}d\omega=-\int_{S(t)}gth^2(t\mathbf{v}-\mathbf{x})\cdot\boldsymbol{\nu}\,dS, \tag{14.50}$$

are valid for the shallow-water-theory equations. These differ from the classical conservation equations.

Example. Higher-order variational problems. For variational problems with Lagrangian functions depending on higher-order derivatives, the main conservation theorems are valid. Only the formula for the c-vector obtained in this case has another form.

Let us consider, in particular, a Lagrangian function depending on second order derivatives,

$$\mathfrak{L}=\mathfrak{L}\,(x,u,\partial u,\partial^2 u). \tag{14.51}$$

The Euler equations in this case are of the form

$$\frac{\delta\mathfrak{L}}{\delta u^k}\equiv\frac{\partial\mathfrak{L}}{\partial u^k}-D_i\frac{\partial\mathfrak{L}}{\partial u_i^k}+D_iD_j\frac{\partial\mathfrak{L}}{\partial u_{ij}^k}=0,\qquad k=1,\cdots,m. \tag{14.52}$$

The invariance criterion of the variational integral with the Lagrangian function (14.51) will have the form

$$\begin{aligned}\mathfrak{L}D_i\xi^i+\xi^i\frac{\partial\mathfrak{L}}{\partial x^i}+\eta^\alpha\frac{\partial\mathfrak{L}}{\partial u^\alpha}+\left(D_i\eta^\alpha-u_j^\alpha D_i\xi^j\right)\frac{\partial\mathfrak{L}}{\partial u_i^\alpha}&\\+\left(D_iD_j\eta^\alpha-u_{jp}^\alpha D_i\xi^p-u_{ip}^\alpha D_j\xi^p-u_p^\alpha D_iD_j\xi^p\right)\frac{\partial\mathfrak{L}}{\partial u_{ij}^\alpha}&=0.\end{aligned} \tag{14.53}$$

The left side of (14.53) can be rewritten in the form

$$\left(\eta^\alpha-u_j^k\xi^j\right)\frac{\partial\mathfrak{L}}{\partial u^k}+D_kA^i, \tag{14.54}$$

where

$$A^i=\left(\eta^\alpha-u_p^k\xi^p\right)\left(\frac{\partial\mathfrak{L}}{\partial u_i^\alpha}-D_j\frac{\partial\mathfrak{L}}{\partial u_{ij}^\alpha}\right)+D_j\left(\eta^\alpha-u_p^k\xi^p\right)\frac{\partial\mathfrak{L}}{\partial u_{ij}^\alpha}+\mathfrak{L}\xi^i. \tag{14.55}$$

So we have the c-vector (14.55) for every one-parameter group of transformations with the Lie tangent vector (9.6) if the variational integral with the Lagrangian function (14.51) is invariant with respect to this group. This is just the essence of Noether's theorem for invariant variational problems with the Lagrangian functions of the form (14.51). In this case Theorem 14.2, as well as Theorem 14.3 (with a corresponding change in Definition 14.2), is valid with c-vector (14.55) instead of the vector defined by the formula (14.7).

Let us consider, as an example, the well-known fourth-order equation

$$u_{tt}+\Delta^2u=0, \tag{14.56}$$

where

$$\Delta=\frac{\partial^2}{\partial x^2}+\frac{\partial^2}{\partial y^2}.$$

The Lagrangian function is

$$\mathfrak{L} = \tfrac{1}{2}u_t^2 - (\Delta u)^2.$$

Taking the one-parameter group of translations in time, we obtain from formula (14.55) the following conservation equation:

$$\frac{\partial}{\partial t}\{u_t^2 + (\Delta u)^2\} + \operatorname{div}\{2u_t \nabla (\Delta u) - \Delta u \nabla u_t\} = 0. \tag{14.57}$$

Further, the invariance of the variational integral with respect to the one-parameter group of rotations with the tangent vector

$$X = y\frac{\partial}{\partial x} - x\frac{\partial}{\partial y} \tag{14.58}$$

leads to the conservation equation

$$\frac{\partial}{\partial t}(\Theta u_t) + \operatorname{div}\{\Theta \nabla (\Delta u) - \Delta u \nabla \Theta + \mathfrak{L}\xi\} = 0, \tag{14.59}$$

where $\Theta = xu_y - yu_x$ and $\xi = (y, -x)$. The variational integral is invariant under the translations of the axes as well. Setting $x = x^1$, $y = x^2$, we can write the conservation equations associated with the 2-parameter group of translations

$$x'^i = x^i + a^i, \qquad i = 1, 2, \tag{14.60}$$

in the form

$$\frac{\partial \tau^i}{\partial t} + \frac{\partial \gamma^{ij}}{\partial x^j} = 0, \qquad i = 1, 2, \tag{14.61}$$

where τ and γ are given by

$$\tau^i = u_t u_i, \qquad \gamma^{ij} = u_i \Delta u_j - u_{ij}\Delta u, \qquad i, j = 1, 2.$$

Recently a more general result has been found which applies to arbitrary systems of partial differential equations. It is based on the notation of a weak Lagrangian (Ibragimov [6], [7]) and the concept of a group of Lie-Bäcklund transformations.

§14.3 Weak Lagrangian Structure

DEFINITION 14.4. A function $\mathfrak{L}$ is called a weak Lagrangian of the system (ω) if

$$\left.\frac{\delta \mathfrak{L}}{\delta u^\alpha}\right|_\Omega = 0, \qquad \alpha = 1, \cdots, m. \tag{14.62}$$

A weak Lagrangian exists for any system (ω). Indeed, it is possible to take for the weak Lagrangian

$$\mathfrak{L}=\mu(x)+\sum_{\nu=1}^{m} a_\nu(x)\omega_\nu^2 \tag{14.63}$$

with arbitrary smooth functions $\mu(x)$, $a_\nu(x)$.

Suppose $\mathfrak{L}$ is a weak Lagrangian for the system (ω) and the corresponding integral (14.3) exists for all the solutions of (ω). Further, let the system (ω) be invariant with respect to a group G of Lie-Bäcklund tangent transformations. Now we can consider the values of the integral of the solutions of the system (ω) and formulate the notion of the invariance of these values with respect to the group G. Here this notion is introduced by using the infinitesimal criteria of the invariance of (ω).

DEFINITION 14.5. A weak Lagrangian $\mathfrak{L}$ of the system (ω) is called relatively G-invariant, where G is generated by a Lie-Bäcklund operator (11.14), if

$$\left(\mathfrak{L}D_i(\xi^i)+\xi^i\frac{\partial\mathfrak{L}}{\partial x^i}+\eta^\alpha\frac{\partial\mathfrak{L}}{\partial u^\alpha}+\zeta_i^\alpha\frac{\partial\mathfrak{L}}{\partial u_i^\alpha}+\cdots\right)\bigg|_\Omega=0. \tag{14.64}$$

Now the basic theorem about conservation laws can be formulated as the following.

THEOREM 14.4. *The conservation law* (14.2) *for the system* (ω) *is valid if and only if* (ω) *is invariant with respect to a Lie-Bäcklund group G and there exists a relatively G-invariant Weak Lagrangian* $\mathfrak{L}$ *of the system* (ω). *Moreover, the conserved vector A is calculated by the following formula*:

$$\begin{aligned}A^i=\mathfrak{L}\xi^i&+(\eta^\alpha-\xi^l u_l^\alpha)\left(\frac{\partial\mathfrak{L}}{\partial u_i^\alpha}+\sum_{s=1}^{\infty}(-1)^s D_{j_1}\cdots D_{j_s}\frac{\partial\mathfrak{L}}{\partial u_{ij_1,\cdots,j_s}^\alpha}\right)\\&+\sum_{r=1}^{\infty}D_{k_1}\cdots D_{k_r}(\eta^\alpha-\xi^l u_l^\alpha)\\&\left(\frac{\partial\mathfrak{L}}{\partial u_{ik_1,\cdots,k_r}^\alpha}+\sum_{s=1}^{\infty}(-1)^s D_{j_1}\cdots D_{j_s}\frac{\partial\mathfrak{L}}{\partial u_{ij_1,\cdots,j_sk_1,\cdots,k_r}^\alpha}\right).\end{aligned} \tag{14.65}$$

For the proof of this theorem see Ibragimov [7].

Lie-Bäcklund groups admitted by a given system of differential equations can be used to obtain new conservation laws for this system from known conservation laws. Namely, if A^i is a conserved vector for the system (ω) and a Lie-Bäcklund operator

$$X=\eta^\alpha(x,u,u,\cdots)\frac{\partial}{\partial u^\alpha}+\cdots$$

is admitted by (ω), then

$$B^i=XA^i \tag{14.66}$$

is also a conserved vector for (ω). This follows directly from the identity (13.6).

§15. Lie via Lie-Bäcklund for Ordinary Differential Equations

The first question we address here is: What is the connection between invariance transformation groups for a given nth-order ordinary differential equation and the equivalent system of first-order ordinary differential equations? In §12, we discussed this question for a second-order ordinary differential equation. There we found, for that case, that the group of point transformations of the original second-order ordinary differential equation is a subgroup of the group which is the image of the group of point transformations admitted by the equivalent system. Moreover, this image is a Lie-Bäcklund tangent transformation group for the original equation. So the notion of Lie-Bäcklund transformations clarifies, in this case and in general, the connection between the group properties of the equivalent system of differential equations and the original system. The importance of the notion of Lie-Bäcklund transformations in this matter is connected with the fact that the equivalence transformation between these two systems is itself a type of Lie-Bäcklund transformation and this equivalence transformation establishes a correspondence between the Lie-Bäcklund groups admitted by the two systems. In the previous case of the second-order ordinary differential equation, this equivalence transformation involves only first-order derivatives. As a result the corresponding image of the infinitesimal operator of the group of point transformations admitted by the equivalent system of first-order equations depends only on first derivatives of the original system, and as a result of this fact, it was proved there that this image is equivalent to a group of Lie tangent transformations on the solutions of the original second-order equation. In this section we analyze the analogous construction for the nth-order ordinary differential equation

$$\frac{d^n u}{dx^n} = 0, \qquad n \geqq 2, \tag{15.1}$$

and establish the equivalence of the Lie point-transformation group admitted by the equivalent first-order system to the Lie-Bäcklund transformation group admitted by (15.1). We then pass to a second question: What is the connection between the various transformation groups admitted by the members of the class of all n-dimensional globally integrable dynamical systems? A construction shows that all members of such a class are group-theoretically equivalent (locally). It follows from the answers to these two questions that any such system has an invariance group isomorphic to one admitted by (15.1). Finally we conclude this section by establishing the existence of a common transitive group admitted by all even-dimensional globally integrable dynamical systems of the same dimensionality, and we present two concrete realizations of this group.

Instead of calculating the most general point-transformation group admitted by the equivalent system of first-order ordinary differential equations as we did for the case $n=2$, we shall (as we could have done in that case) directly determine the most general Lie-Bäcklund operator admitted by the original nth-order system. (It can be shown that both ways give exactly the same results.) So we turn to the problem of finding the most general Lie-Bäcklund operator admitted by (15.1). In the notation of §13, if the nth-order ordinary differential equation (15.1),

$$\underset{n}{u} = 0, \tag{15.2}$$

admits a Lie-Bäcklund operator

$$X = \eta\left(x, u, \underset{1}{u}, \cdots, \underset{n-1}{u}\right)\frac{\partial}{\partial u} + \cdots \tag{15.3}$$

then the invariance of (15.2) under X given by (15.3) is expressed by

$$D_x^n \eta|_{(15.2)} = 0. \tag{15.4}$$

First, we find the invariants of the operator

$$\underset{n-2}{D}{}_x = \frac{\partial}{\partial x} + \underset{1}{u}\frac{\partial}{\partial u} + \cdots + \underset{n-1}{u}\frac{\partial}{\partial \underset{n-2}{u}}. \tag{15.5}$$

The characteristic equations for the operator (15.5) are given by

$$\frac{dx}{1} = \frac{du}{\underset{1}{u}} = \frac{d\underset{1}{u}}{\underset{2}{u}} = \cdots = \frac{d\underset{j}{u}}{\underset{j-1}{u}} = \cdots = \frac{d\underset{n-3}{u}}{\underset{n-2}{u}} = \frac{d\underset{n-2}{u}}{\underset{n-1}{u}}, \tag{15.6}$$

which implies that the invariants of (15.5) are given by

$$C_{n-k} = \sum_{j=1}^{k} (-1)^{k-j} \underset{n-j}{u} \frac{x^{k-j}}{(k-j)!}, \tag{15.7}$$

where $k = 1, 2, \cdots, n$. Now employing the same techniques that were used in §12 to solve (12.31), it follows that the general solution of (15.4) can be expressed in the form

$$\eta = \sum_{j=0}^{n-1} \frac{x^j}{j!} g_j(c_0, \cdots, c_{n-1}). \tag{15.8}$$

In general, these operators cannot be Lie tangent operators, because of their dependence on $\underset{k}{u}$, $k > 1$; but the Lie-Bäcklund transformation of (15.1) which yields the equivalent first-order system when applied to (15.3) with η given by (15.8) induces an equivalent set of Lie operators admitted by the equivalent first-order system. In the case $n = 2$, we recover our previous result given by (12.23). Further, in this case the Lie-Bäcklund operator corresponding to η given by (15.8) is equivalent to a Lie tangent operator on the solutions of (12.17). Indeed,

$$\eta|_{(12.17)} = g_0\left(u - x\underset{1}{u}, \underset{1}{u}\right) + ug_1\left(u - x\underset{1}{u}, \underset{1}{u}\right), \tag{15.9}$$

which agrees with (12.36) in another form. Finally, before turning to the second question, note that the maximal group of Lie point transformations admitted by (15.1) for $n>2$ is generated by the $n+4$ infinitesimal operators (Lie [6] p. 298)

$$\frac{\partial}{\partial x},\quad \frac{\partial}{\partial u},\quad x\frac{\partial}{\partial x},\quad u\frac{\partial}{\partial u},\quad x\frac{\partial}{\partial u},\quad x^2\frac{\partial}{\partial u},\cdots,\quad x^{n-1}\frac{\partial}{\partial u},$$
$$x^2\frac{\partial}{\partial x}+(n-1)xu\frac{\partial}{\partial u}.$$

In a more general context, consider two dynamical systems, i.e., the following Lie equations for m dependent variables:

$$\frac{du^\alpha}{dx}=\Phi^\alpha(u),\qquad u|_{x=0}=u_0, \tag{15.10}$$

$$\frac{dv^\alpha}{dx}=\Psi^\alpha(u),\qquad v|_{x=0}=v_0, \tag{15.11}$$

together with the Lie groups defined by (15.10), (15.11) with group parameter x, which act as transformation groups on the phase space (u-space). In addition to these groups of transformations in the phase space, consider the induced groups of transformations in the (x,u)-space with group parameter τ defined by the following Lie equations:

$$\begin{aligned}\frac{dx}{d\tau}&=1,\\ \frac{du}{d\tau}&=\Phi^\alpha(u),\end{aligned} \tag{15.12}$$

with $x|_{\tau=0}=0$, $u|_{\tau=0}=u_0$, and

$$\begin{aligned}\frac{dx}{d\tau}&=1,\\ \frac{dv^\alpha}{d\tau}&=\Psi^\alpha(u),\end{aligned} \tag{15.13}$$

with $x|_{\tau=0}=0$, $v|_{\tau=0}=v_0$. Suppose that (15.12) and (15.13) are globally integrable and the corresponding global Lie groups acting in the (x, u)-space are given by

$$\begin{aligned}x_\tau&=\tau,\\ u_\tau&=F_1(u_0;\,\tau),\end{aligned} \tag{15.14}$$

and

$$\begin{aligned}x_\tau&=\tau,\\ v_\tau&=F_2(v_0;\,\tau).\end{aligned} \tag{15.15}$$

Equations (15.14), (15.15) permit us to establish a one-to-one correspondence between the graphs of the solutions of (15.10) and (15.11). This correspondence is global and is given by

$$\begin{aligned} &x = x, \\ &u_x^\alpha = F_1^\alpha(F_2(v_x; -x); x), \qquad \alpha = 1, \cdots, m, \end{aligned} \tag{15.16}$$

and its inverse

$$\begin{aligned} &x = x, \\ &v^\alpha = F_2^\alpha(F_1(u_x; -x); x), \qquad \alpha = 1, \cdots, m. \end{aligned} \tag{15.17}$$

This is pictorially illustrated by Figs. 6 and 7, which are described in Example 15.1 below. We can see that if one of the equations, say (15.12), is linear, while the other one, (15.13), is nonlinear, then the map (15.16) globally linearizes (15.13). Here we shall employ these maps to exhibit the group equivalence of the dynamical systems (15.10), (15.11).

Suppose that (15.12) admits an invariance transformation group G_1 with group parameters a which can be a group of Lie point transformations, Lie tangent transformations, or Lie-Bäcklund tangent transformations

$$G_1: \begin{cases} x' = f_1\left(x, u, \underset{1}{u}, \ldots; a\right), \\ u' = \phi_1\left(x, u, \underset{1}{u}, \ldots; a\right), \\ \cdots\cdots\cdots\cdots \end{cases} \tag{15.18}$$

The one-to-one correspondence (15.16), (15.17) converts the invariance group G_1 into the isomorphic group G_2 which is admitted by equation (15.13) and is given by

$$G_2: \begin{cases} x' = f_1(x, F_1(F_2(v; -x); x), \cdots; a) \overset{\text{Def}}{=} f_2(x, v, \underset{1}{v}, \cdots; a), \\ v' = F_2\Big(F_1\big(\phi_1(x, F_1(F_2(v; -x); x), \cdots; a); \\ \qquad -f_2\big(x, v, \underset{1}{v}, \cdots; a\big)\big); f_2\big(x, v, \underset{1}{v}, \cdots\big)\Big) \\ \qquad \overset{\text{Def}}{=} \phi_2(x, v, \underset{1}{v}, \ldots; a), \\ \cdots\cdots\cdots\cdots\cdots\cdots\cdots\cdots \end{cases} \tag{15.19}$$

Note that although G_1 may be a group of Lie point transformations, in general G_2 will be a group of Lie-Bäcklund transformations. Equations (15.16), (15.17), and (15.19) with (15.18) given by a Lie point transformation were established for even-dimensional globally integrable systems in Anderson and Peterson [1].

Now given any invariance transformation group G_0 admitted by (15.1) for a fixed n, then G_0 induces an equivalent invariance transformation group G_1 for the equivalent first-order system (15.12). Therefore the map (15.19) induces an isomorphic invariance group for any other integrable system (15.13) of the same dimensionality. Hence we have established the group equivalence (local) of all members of the class of n-dimensional globally integrable dynamical systems. Further, the results obtained thus far directly imply that every n-dimensional globally integrable dynamical system has an invariance group isomorphic to one admitted by (15.1).

Before we proceed to more particular developments and concrete realizations, we shall quickly extend these results to the case where the global solution of the system (15.11) may be known only in closed form for a reparametrization of the usual time parametrization. This extension will be needed later in this section in order to utilize such a parametrized global solution directly in an application of these developments to the one-body Keplerian problem (see Example 15.2). This extension is accomplished by first replacing (15.13) with the reparametrized Lie equation

$$\begin{aligned}\frac{dx}{d\psi} &\overset{\text{Def}}{=} \Xi^0(v),\\ \frac{dv^\alpha}{d\psi} &= \Psi^\alpha(v)\,\Xi^0(v) \overset{\text{Def}}{=} \Xi^\alpha(v), \qquad \alpha=1,\cdots,m,\end{aligned}\tag{15.20}$$

with $x|_{\psi=0}=0$, $v|_{\psi=0}=v_0$, and replacing (15.15) by

$$\begin{aligned}x_\psi &= F_2^0(v_0;\,\psi) \overset{\text{Def}}{=} h_{v_0}(\psi),\\ v_\psi^\alpha &= F_2^\alpha(v_0;\,\psi), \qquad \alpha=1,\cdots,m.\end{aligned}\tag{15.21}$$

Then (15.16), (15.17) become

$$\begin{aligned}x &= h_{v_0}(\psi)\\ u_{x_\psi}^\alpha &= F_1^\alpha\big(F_2(v_\psi;\,-\psi);\,h_{v_0}(\psi)\big), \qquad \alpha=1,\cdots,m,\end{aligned}\tag{15.22}$$

with inverse

$$\begin{aligned}x &= h_{v_0}(\psi)\\ v_x^\alpha &= F_2^\alpha\big(F_1\big(u_\psi;\,-h_{v_0}(\psi)\big);\,\psi\big), \qquad \alpha=1,\cdots,m,\end{aligned}\tag{15.23}$$

where we have freely interchanged the subscripts x, τ, ψ on u and/or v whenever it is understood that $x=\tau$ and/or $x=h_{v_0}(\psi)$. Note that the necessity of the development presented in this section stems from the well-known fact that, in general, $\psi=h_{v_0}^{-1}(x)$ does not exist in closed form—e.g., in the Kepler problem discussed in Example 15.2. This extension is completed with the statement that the one-to-one correspondence (15.22), (15.23) converts the invariance group G_1 into the isomorphic group G_2 admitted by (15.11), where G_2 is given by

$$
\begin{aligned}
& x' = f_1\big(x, F_1\big(F_2\big(v; -h_{v_0}^{-1}(x)\big); x\big), \ldots; a\big) \overset{\text{Def}}{=} f_2(x, v, \ldots; a),\\
(15.24)\qquad & v' = F_2\big(F_1\big(\phi_1\big(x, F_1\big(F_2\big(v; -h_{v_0}^{-1}(x)\big); x\big), \ldots; a\big);\\
& \quad -f_2(x, v, \ldots; a)\big); h_{v_0'}^{-1}(f_2(x, v, \ldots; a)\big)\\
& \quad \overset{\text{Def}}{=} \phi_2(x, v, \ldots; a),\\
& \cdots\cdots\cdots\cdots\cdots\cdots
\end{aligned}
$$

where the connection between x and ψ is given by (15.21), and by definition $v_0' = v'|_{x'=0}$. Equation (15.24) replaces (15.19).

Now, turning to more particular results, we can obtain a concrete realization of G_1 and hence G_2 for all completely integrable even-dimensional dynamical systems. In particular, for all globally integrable $2n$-dimensional dynamical systems there exists a transitive invariance transformation group isomorphic to the group of special (determinant $=1$) linear transformations in $n+2$ real dimensions [SL $(n+2, \mathbb{R})$]. This is obtained by first taking for the system (15.10) Hamilton's equations of motion for a free particle with mass $\overline{m}$ and described by $2n$ canonical variables [here, n Euclidean-space degrees of freedom $(u^1, \cdots, u^n)$ and the n associated canonical momenta $(u^{n+1}, \cdots, u^{2n})$] in the form

$$
(15.25)\qquad
\begin{aligned}
\frac{du^i}{dx} &= \frac{u^{n+i}}{\overline{m}}, \qquad & i = 1, \cdots, n,\\
\frac{du^{n+i}}{dx} &= 0, \qquad & i = 1, \cdots, n.
\end{aligned}
$$

The connection between the usual t-x-p (time–space–linear-momentum) notation and the x-u notation employed in (15.10) is $x = t$, $u = (u^1, \cdots, u^n, u^{n+1}, \cdots, u^{2n}) = (x^1, \cdots, x^n, p^1, \cdots, p^n)$. The corresponding global transformation group is

$$
(15.26)\qquad
\begin{aligned}
u^i &= F_1^i(u_0; x) = u_0^i + \frac{u_0^{n+i}}{\overline{m}} x, \qquad & i = 1, \cdots, n,\\
u^{n+i} &= F_1^{n+i}(u_0; x) = u_0^{n+i}, \qquad & i = 1, \cdots, n.
\end{aligned}
$$

For G_1 [equation (15.18)] we take the maximal space-time Lie invariance transformation group admitted by a Newtonian free particle system with n degrees of freedom (Anderson, di Franco, and Smith [1]). This is a transitive Lie point transformation group for the space-time coordinates $(u^1,\cdots,u^n)-x$ extended to the canonical momenta $(u^{n+1},\cdots,u^{2n})$ and is explicitly realized in the form

$$x'=\frac{a_{00}x+a_{0s}u^s+a_{0,\,n+1}}{a_{n+1,\,0}x+a_{n+1,\,b}u^b+a_{n+1,\,n+1}}, \tag{15.27a}$$

$$u'^i=\frac{a_{i0}x+a_{is}u^s+a_{i,\,n+1}}{a_{n+1,\,0}x+a_{n+1,\,b}u^b+a_{n+1,\,n+1}}, \tag{15.27b}$$

$$u'^{n+i}=\bar{m}\frac{\begin{aligned}&a_{n+1,\,0}a_{is}L_{0s}+a_{n+1,\,s}a_{i0}L_{s0}\\&\quad+a_{n+1,\,0}a_{i,\,n+1}L_{0,\,n+1}+a_{n+1,\,n+1}a_{i0}L_{n+1,\,0}\\&\quad+a_{n+1,\,r}a_{is}L_{rs}\\&\quad+a_{n+1,\,n+1}a_{is}L_{n+1,\,s}+a_{n+1,\,s}a_{i,\,n+1}L_{s,\,n+1}\end{aligned}}{\begin{aligned}&a_{n+1,\,0}a_{0c}L_{0c}+a_{n+1,\,c}a_{00}L_{c0}\\&\quad+a_{n+1,\,0}a_{0,\,n+1}L_{0,\,n+1}+a_{n+1,\,n+1}a_{00}L_{n+1,\,0}\\&\quad+a_{n+1,\,b}a_{0c}L_{bc}\\&\quad+a_{n+1,\,n+1}a_{0c}L_{n+1,\,c}+a_{n+1,\,c}a_{0,\,n+1}L_{c,\,n+1}\end{aligned}} \tag{15.27c}$$

where

$$\begin{aligned}L_{0s}&=xu^{n+s}-\bar{m}u^s=-L_{s0},\\L_{rs}&=u^ru^{n+s}-u^su^{n+r},\\L_{n+1,\,s}&=u^{n+s}=-L_{s,\,n+1},\\L_{n+1,\,0}&=\bar{m}=-L_{0,\,n+1},\end{aligned}$$

and all free indices take on the values $1,\cdots,n$ and all dummy indices are summed over these values. The transformation group G_1 given by (15.27) is a nonlinear faithful representation of the quotient group of the general linear group in $n+2$ real dimensions by the subgroup of all nonzero multipliers of the identity (e.g., Birkhoff and MacLane [1]). Equivalently, (15.27) is isomorphic to the special linear group in $n+2$ real dimensions [SL $(n+2,\mathbb{R})$], i.e.,

$$\det\begin{Vmatrix}a_{00}&a_{01}&\cdots&a_{0n}&a_{0,\,n+1}\\\vdots&\vdots&&\vdots&\vdots\\a_{n+1,\,0}&a_{n+1,\,1}&\cdots&a_{n+1,\,n}&a_{n+1,\,n+1}\end{Vmatrix}=1.$$

The group (15.27) is the maximal space-time Lie transformation group admitted by (15.25), in the sense that it contains all the local one-parameter Lie transformation groups corresponding to all the linearly independent Lie operators of the form

$$X=\xi(x,u^1,\cdots,u^n)\frac{\partial}{\partial x}+\eta^i(x,u^1,\cdots,u^n)\frac{\partial}{\partial v^i}$$
$$+\eta^{n+i}(x,u^1,\cdots,u^{2n})\frac{\partial}{\partial v^{n+i}},$$

where $u^1,\cdots,u^n$ are required to correspond to n Euclidean space coordinates. If for each parameter $a_{\beta\alpha}$ (α, $\beta=0,1,\cdots,n,n+1$) we introduce the Lie operator $X_{\alpha\beta}$ of the associated one-parameter Lie transformation group according to the usual convention [see e.g. (9.6)], then

$$X_{\alpha\beta}=\xi_{\alpha\beta}\frac{\partial}{\partial x}+\eta^i_{\alpha\beta}\frac{\partial}{\partial u^i}+\eta^{n+i}_{\alpha\beta}\frac{\partial}{\partial u^{n+i}}, \tag{15.28}$$

where

$$X_{00}=x\frac{\partial}{\partial x}-u^{n+i}\frac{\partial}{\partial u^{n+i}},$$
$$X_{i0}=u^i\frac{\partial}{\partial x}-\frac{u^{n+i}u^{n+j}}{\overline{m}}\frac{\partial}{\partial u^{n+j}},$$
$$X_{0i}=x\frac{\partial}{\partial u^i}+\overline{m}\frac{\partial}{\partial u^{n+i}},$$
$$X_{ij}=u^i\frac{\partial}{\partial u^j}+u^{n+i}\frac{\partial}{\partial u^{n+j}},$$
$$X_{n+1,0}=\frac{\partial}{\partial x},$$
$$X_{0,n+1}=-x^2\frac{\partial}{\partial x}-xu^j\frac{\partial}{\partial u^j}+(-\overline{m}u^i+xu^{n+i})\frac{\partial}{\partial u^{n+i}},$$
$$X_{n+1,i}=\frac{\partial}{\partial u^i},$$
$$X_{i,n+1}=-xu^i\frac{\partial}{\partial x}-u^iu^j\frac{\partial}{\partial u^j}+\frac{u^{n+i}}{\overline{m}}(-\overline{m}u^j+xu^{n+j})\frac{\partial}{\partial u^{n+j}},$$
$$X_{n+1,n+1}=-X_{00}-X_{11}-\cdots-X_{nn}.$$

The $X_{\alpha\beta}$'s satisfy the commutation relations of the Lie algebra of the general linear group in $n+2$ real dimensions,

$$[X_{\alpha\beta},X_{\sigma\rho}]=\delta_{\beta\sigma}X_{\alpha\rho}-\delta_{\alpha\rho}X_{\sigma\beta}, \tag{15.29}$$

and any $(n+2)^2-1$ linearly independent $X_{\alpha\beta}$'s constitute a basis for the Lie algebra of the corresponding special linear group.

The identification of the system (15.25) with the system (15.10) and the identification of the transformation group (15.27) with the group G_1 given by (15.18) yields for each system of the type (15.11) the isomorphic transitive invariance transformation group G_2 corresponding to (15.24) in the explicit form (Anderson and Merner [1])

$$x' = \frac{a_{00}x + a_{0s}A^s + a_{0,n+1}}{a_{n+1,0}x + a_{n+1,b}A^b + a_{n+1,n+1}} = f_2(x, v; a), \tag{15.30a}$$

$$v'^{\alpha} = F_2^{\alpha}\left(B^i - \frac{f_2(x, v; a)}{\overline{m}} B^{n+i}, B^{n+j}; h_{v_0'}^{-1}(f_2(x, v; a))\right), \qquad \alpha = 1, \cdots, 2n, \tag{15.30b}$$

where

$$A^i = F_2^i\left(v; -h_{v_0}^{-1}(x)\right) + \frac{x}{\overline{m}} F_2^{n+i}\left(v; -h_{v_0}^{-1}(x)\right),$$

$$A^{n+i} = F_2^{n+i}\left(v; -h_{v_0}^{-1}(x)\right),$$

$$B^i = \frac{a_{i0}x + a_{is}A^s + A_{i,n+1}}{a_{n+1,0}x + a_{n+1,b}A^b + a_{n+1,n+1}},$$

$$B^{n+i} = \overline{m}\frac{\begin{array}{l} a_{n+1,0}a_{is}\overline{L}_{0s} + a_{n+1,s}a_{i0}\overline{L}_{s0} \\ \quad + a_{n+1,0}a_{i,n+1}\overline{L}_{0,n+1} + a_{n+1,n+1}a_{i0}\overline{L}_{n+1,0} \\ \quad + a_{n+1,r}a_{is}\overline{L}_{rs} \\ \quad + a_{n+1,n+1}a_{is}\overline{L}_{n+1,s} + a_{n+1,s}a_{i,n+1}\overline{L}_{s,n+1} \end{array}}{\begin{array}{l} a_{n+1,0}a_{0c}\overline{L}_{0c} + a_{n+1,c}a_{00}\overline{L}_{c0} \\ \quad + a_{n+1,0}a_{0,n+1}\overline{L}_{0,n+1} + a_{n+1,n+1}a_{00}\overline{L}_{n+1,0} \\ \quad + a_{n+1,b}a_{0c}\overline{L}_{bc} \\ \quad + a_{n+1,n+1}a_{0c}\overline{L}_{n+1,c} + a_{n+1,c}a_{0,n+1}\overline{L}_{c,n+1} \end{array}}$$

and

$$\overline{L}_{0s} = xA^{n+s} - \overline{m}A^s = -\overline{L}_{s0},$$

$$\overline{L}_{rs} = A^rA^{n+s} - A^sA^{n+r},$$

$$\overline{L}_{n+1,s} = A^{n+s} = -\overline{L}_{s,n+1},$$

$$\overline{L}_{n+1,0} = \overline{m} = -\overline{L}_{0,n+1}.$$

Let $Y_{\alpha\beta}$ denote the induced Lie algebraic element corresponding to $X_{\alpha\beta}$ given by (15.28). Then the induced Lie algebraic structure follows in the usual way [see e.g. (9.6)], or equivalently it is given by

$$Y_{\alpha\beta}=(X_{\alpha\beta}x)\big|_{(15.22)}\frac{\partial}{\partial x}+\left(X_{\alpha\beta}F_2^{\gamma}\left(F_1(u;\,-x);\,h_{v_o}^{-1}(x)\right)\right)\Big|_{(15.22)}\frac{\partial}{\partial u^{\gamma}}$$

where the $Y_{\alpha\beta}$'s automatically obey the commutation relations (15.29).

Example 15.1. *Anharmonic Oscillator* (Anderson and Peterson [1]). An application of the map (15.19) yields an induced transitive SL(3, R) Lie invariance transformation group for the anharmonic oscillator system with one Euclidean space degree of freedom v^1, conjugate momentum v^2, and mass m corresponding to the dynamical system

$$\text{(15.31)}\qquad \begin{aligned}\frac{dv^1}{dx}&=\frac{v^2}{m},\\ \frac{dv^2}{dx}&=-K\,(v^1)^3,\end{aligned}$$

The well-known general solution to this problem, expressed in the form (15.15), is given by

$$\text{(15.32a)}\qquad t'=t+\tau,$$

$$\text{(15.32b)}\qquad v'^1=\frac{(v^2/m\omega)\,\mathrm{sn}\,(\omega\tau,i)+v^1\,\mathrm{cn}\,(\omega\tau,i)\,\mathrm{dn}\,(\omega\tau,i)}{1+(v^1/v_M)^2\,\mathrm{sn}^2(\omega\tau,i)}=F_2^1(u;\tau),$$

$$\text{(15.32c)}\qquad v'^2=\frac{v^2\,\mathrm{cn}\,(\omega\tau,i)\,\mathrm{dn}\,(\omega\tau,i)\left[1-(v^1/v_M)^2\,\mathrm{sn}^2(\omega\tau,i)\right]}{\left[1+(v^1/v_M)^2\,\mathrm{sn}^2(\omega\tau,i)\right]^2}-\frac{2m\omega\,(v^1/v_M)\,\mathrm{sn}\,(\omega\tau,i)\left[(v^1/v_M)^2+\mathrm{sn}^2(\omega\tau,i)\right]}{\left[1+(v^1/v_M)^2\,\mathrm{sn}^2(\omega\tau,i)\right]^2}=F_2^2(u;\tau),$$

where

$$\omega=\left(\frac{K}{2m}\right)^{1/2}v_M,\qquad v_M=\left(\frac{4E}{K}\right)^{1/4},$$

$$E=\frac{(v^2)^2}{2m}+\frac{K\,(v^1)^4}{4}=\frac{(v'^2)^2}{2m}+\frac{(Kv'^1)^4}{4},$$

$i=\sqrt{-1}$, and sn, cn, dn are the Jacobian elliptic sine, cosine, and delta functions, respectively. The unprimed quantities are given in terms of the primed ones by (15.32) with an interchange of primed and unprimed quantities and

$x \to -x$. Note these changes must also be made in ω and v_M. Then (15.30) becomes

$$x' = f_2(x, v; a) = \frac{a_{00}x + a_1 A^1(t, u) + a_{02}}{a_{20}x + a_{21}A^1(t, u) + a_{22}}, \tag{15.33a}$$

$$\begin{aligned} v'^1 &= \phi_2^1(x, v; a) \\ &= F_2^1\left(B^1 - \frac{f_2(x, v; a)}{\bar{m}} B^2, B^2; f_2(x, v; a)\right), \end{aligned} \tag{15.33b}$$

$$\begin{aligned} v'^2 &= \phi_2^2(x, v; a) \\ &= F_2^2\left(B^1 - \frac{f_2(x, v; a)}{\bar{m}} B^2, B^2; f_2(x, v; a)\right), \end{aligned} \tag{15.33c}$$

where

$$F_2(v; x) = \left(F_2^1(v^1, v^2; x), F_2^2(v^1, v^2; x)\right)$$

is given by (15.32), and

$$A^1(x, v) = F_2^1(v; -x) + \frac{x}{\bar{m}} F_2^2(v; -x),$$

$$A^2(x, v) = F_2^2(v; -x),$$

$$B^1(x, v) = \frac{a_{10}x + a_{11}A^1(x, v) + a_{12}}{a_{20}x + a_{21}A^1(x, v) + a_{22}},$$

$$B^2(x, v) = \bar{m}\frac{\begin{array}{c}(a_{20}a_{11} - a_{21}a_{10})\bar{L}_{01}(x, v) + (a_{21}a_{12} - a_{22}a_{11})\bar{L}_{12}(x, v) \\ + (a_{20}a_{12} - a_{22}a_{10})\bar{L}_{02}(x, v)\end{array}}{\begin{array}{c}(a_{20}a_{01} - a_{21}a_{00})\bar{L}_{01}(x, v) + (a_{21}a_{02} - a_{22}a_{01})\bar{L}_{12}(x, v) \\ + (a_{20}a_{02} - a_{22}a_{00})\bar{L}_{02}(x, v)\end{array}}$$

with

$$\begin{aligned} \bar{L}_{01}(x, v) &= xA^2(x, v) - \bar{m}A^1(x, v), \\ \bar{L}_{12}(x, v) &= -A^2(x, v), \\ \bar{L}_{02}(x, v) &= -\bar{m}. \end{aligned}$$

Figure 8 depicts two phase curves—one for the free-particle system and one for the anharmonic-oscillator system—while Fig. 9 depicts the associated graphs and a pair of points on these graphs which correspond under the map (15.16) and its inverse (15.17).

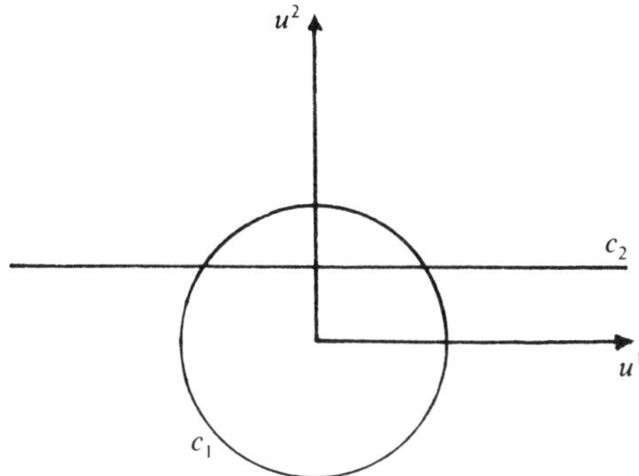

FIG. 8. Two phase curves—the curve C_1 is for the free particle system while C_2 is for the anharmonic oscillator system.

Example 15.2. *One-body Keplerian problem* (Anderson and Merner [1]). An important example of a nonlinear $2n$-dimensional system whose global solution is *only* known in closed form for a reparametrization of the usual time parametrization is provided by the one-body Keplerian problem. This well-known classical problem is posed as the determination of the motion of a mass m moving under the action of a central force inversely proportional to the square of the distance from a fixed center. Here we shall employ the conveniently compact form of the global solution given by Goodyear [1], which is a modification of the form of the solution given by Stumpff [1] and that given later and independently by Herrick [1]. Its utility for the present application lies in the fact that it provides one form which is valid for all the circular, elliptic, parabolic, hyperbolic, and rectilinear orbits of the attractive force, and all the hyperbolic and rectilinear orbits of the repulsive force.

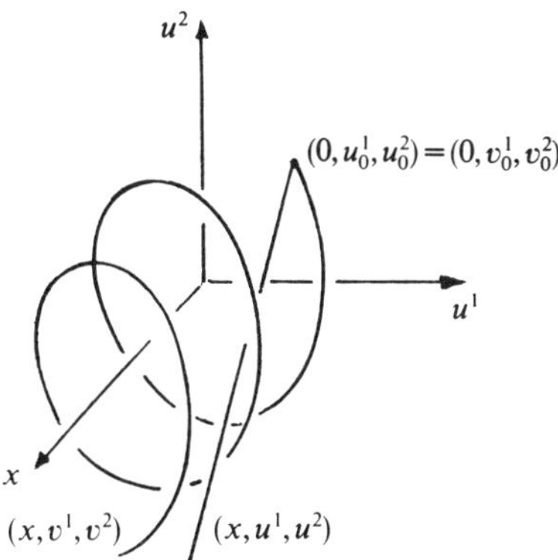

FIG. 9. Graphs associated with the phase curves shown in Fig. 8 and a pair of points on these graphs which correspond to each other under the maps (15.16), (15.17).

Hamilton's equations of motion for the one-body Keplerian problem may be taken in the form

$$\text{(15.34)}\qquad \begin{aligned} \frac{dv^i}{dx} &= \frac{v^{3+i}}{m}, \qquad i=1,2,3,\\ \frac{dv^{3+i}}{dx} &= -\frac{mgv^i}{r^3}, \qquad i=1,2,3, \end{aligned}$$

where

$$r=\left[(v^1)^2+(v^2)^2+(v^3)^2\right]^{1/2},$$

and the "coupling" constant g is positive for an attractive force, negative for a repulsive force, and zero for free particle motion.

Goodyear's parametrized general global solution of the system (15.34) with parameter ψ corresponding to our (15.21) is

$$\text{(15.35)}\qquad \begin{aligned} x &= r_0 s_1 + \sigma_0 s_2 + g s_3,\\ v^i &= v_0^i \alpha + v_0^{3+i}\frac{\beta}{m}, \qquad i=1,2,3,\\ v^{3+i} &= m v_0^i \delta + v_0^{3+i}\gamma, \qquad i=1,2,3, \end{aligned}$$

where

$$\alpha = 1 - g s_2/r_0, \qquad \beta = r_0 s_1 + \sigma_0 s_2,$$

$$\delta = -\frac{g s_1}{r r_0}, \qquad \gamma = 1 - \frac{g s_2}{r}$$

$$r_0 = \left(v_0^i v_0^i\right)^{1/2}, \qquad \sigma_0 = v_0^i v_0^{3+i}/m,$$

$$e = \frac{v_0^{3+i} v_0^{3+i}}{2m^2} - \frac{g}{r_0},$$

$$r = r_0 s_0 + \sigma_0 s_1 + g s_2,$$

and

$$s_0 = \cosh \sqrt{2e}\,\psi,$$

$$s_1 = \frac{1}{2e}\left(\frac{d}{d\psi}\cosh\sqrt{2e}\,\psi\right),$$

$$s_2 = \frac{s_0 - 1}{2e},$$

$$s_3 = \frac{s_1 - \psi}{2e}.$$

(All repeated indices are to be summed over the values 1, 2, 3.) The interested reader is referred to Goodyear's paper for the explicit connection between the parameter ψ and Kepler's eccentric anomaly.

Now, the substitution of (15.35) into the induced SL $(n+2, \mathbb{R})$ group algorithm (15.30) yields for each value of g a 24-parameter transitive transformation group for the one-body Keplerian problem. This group is isomorphic to SL $(5\mathbb{R})$. Note that in the process of obtaining this invariance group, we have made use of (15.22) and (15.33), which, because we have identified (15.25) with (15.10), constitute a global linearization of (15.34).

§16. Group-Theoretical Equivalence of Quantum-Mechanical Systems[1]

Suppose two evolution equations, i.e., the initial-value problems

$$\left.\begin{aligned} &(16.1)\quad \frac{\partial u^\alpha}{\partial t}=\Phi^\alpha\Big(x, u, \underset{1}{u},\cdots,\underset{k}{u}\Big), &&u^\alpha|_{t=0}=u_0^\alpha(x),\\ &(16.2)\quad \frac{\partial v^\alpha}{\partial t}=\psi^\alpha\Big(x, v, \underset{1}{v},\cdots,\underset{k}{v}\Big), &&v^\alpha|_{t=0}=v_0^\alpha(x), \end{aligned}\right\}\quad \alpha=1,\cdots,m,$$

have the solutions $u_t^\alpha(x)=K_1^\alpha(u_0(x);\ t)$, $v_t^\alpha(x)=K_2^\alpha(v_0(x);\ t)$, respectively, which are analytic in t and x. Then these solutions satisfy the group property

$$K_i^\alpha(w_0(x), t+s)=K_i^\alpha(w_t(x); s)=K_i^\alpha(w_s(x); t), \tag{16.3}$$

for $t, s\in I\subset\mathbb{R}$, where I is an open interval containing zero. The group property (16.3) allows one to establish locally, in a space $\mathcal{F}_x$ of x and functions $u(x)$, a one-to-one correspondence between the t-parametrized solutions $u_t(x)$, $v_t(x)$ for (16.1), (16.2), respectively, which evolve from the same initial condition $u_0(x)=v_0(x)$. But, even if the $K_i\,(i=1, 2)$ are global solutions, this correspondence is still only local in general (e.g., in the case that one of the solutions, say K_1, is periodic and the other, K_2, is not). This latter restriction (in general, only local one-to-one correspondence) can be removed if we introduce instead of the space $\mathcal{F}_x$, the augmented space $\mathcal{F}_{x,\,t}$ of t, x, and functions $u_t(x)$. We can then pass for the system described by (16.1) to a group in the space $\mathcal{F}_{x,\,t}$ described by the evolution system

$$\begin{aligned} &\frac{\partial t}{\partial \tau}=1, && t|_{\tau=0}=0,\\ &\frac{\partial x}{\partial \tau}=0, && x|_{\tau=0}=x,\\ &\frac{\partial u}{\partial \tau}=\Phi\left(x, u, \underset{1}{u},\cdots,\underset{k}{u}\right), && u|_{\tau=0}=u_0(x), \end{aligned} \tag{16.4}$$

[1]See Anderson, Barut, and Ibragimov [1].

with solution

$$
\begin{aligned}
&t_\tau = \tau, \\
(16.5)\qquad &x_\tau = x, \\
&u_\tau(x) = K_1(u_0(x); \tau).
\end{aligned}
$$

and similarly for the system described by (16.2). Now, if the $K_i (i=1, 2)$ are global solutions, one can always establish a global one-to-one correspondence between the elements of the space $\mathfrak{F}_{x,t}$. Note that if instead of the autonomous systems (16.1), (16.2), one started with nonautonomous systems, then one could pass to the systems (16.1), (16.2) by reparametrizing in terms of an additional coordinate and then parallel the preceding argument.

The preceding transformations can be used to establish a connection between the Lie-Bäcklund groups admitted by (16.1) and (16.2). In order to do this it is necessary to recast (16.4) and (16.5) as a Lie-Bäcklund equation and the corresponding Lie-Bäcklund group of transformations, respectively. In particular, we associate with (16.1), (16.2) the equivalent induced Lie-Bäcklund groups defined by the Lie-Bäcklund equations

$$
\begin{aligned}
&\frac{dt'}{d\tau} = 1, \qquad t'|_{\tau=0} = 0, \\
&\frac{dx'}{d\tau} = 0, \qquad x'|_{\tau=0} = x, \\
(16.6)\qquad &\frac{du'^\alpha}{d\tau} = \Phi^\alpha\Big(x', u', \underset{1}{u}', \cdots, \underset{k}{u}'\Big), \qquad u'^\alpha|_{\tau=0} = u_0^\alpha(x), \\
&\frac{du_i'^\alpha}{d\tau} = D_i\Phi^\alpha\Big(x', u', \underset{1}{u}', \cdots, \underset{k}{u}'\Big), \qquad u_i'^\alpha|_{\tau=0} = \frac{\partial}{\partial x^i} u_0^\alpha(x), \\
&\frac{du_{ii_2}'^\alpha}{d\tau} = D_{i_1}D_{i_2}\Phi^\alpha\Big(x', u', \underset{1}{u}', \cdots, \underset{k}{u}'\Big), \qquad u_{i_1 i_2}'^\alpha|_{\tau=0} = \frac{\partial}{\partial x_1^i}\frac{\partial}{\partial x_2^i} u_0^\alpha(x), \\
&\cdots\cdots\cdots\cdots\cdots\cdots\cdots\cdots\cdots\cdots
\end{aligned}
$$

and

$$
\begin{aligned}
&\frac{dt'}{d\tau} = 1, \qquad t'|_{\tau=0} = 0, \\
&\frac{dx'}{d\tau} = 0, \qquad x'|_{\tau=0} = x, \\
(16.7)\qquad &\frac{dv'^\alpha}{d\tau} = \Psi^\alpha(x', v', \underset{1}{v}', \cdots, \underset{k}{v}'), \qquad v'|_{\tau=0} = v_0^\alpha(x), \\
&\frac{dv_i'^\alpha}{d\tau} = D_i\Psi^\alpha(x', v', \underset{1}{v}', \cdots, \underset{k}{v}'), \qquad v_i'^\alpha|_{\tau=0} = \frac{\partial}{\partial x^i} v_0^\alpha(x), \\
&\frac{dv_{i_1 i_2}^\alpha}{d\tau} = D_{i_1}D_{i_2}\Psi^\alpha(x', v', \underset{1}{v}', \cdots, \underset{k}{v}'), \qquad v_{i_1 i_2}'^\alpha|_{\tau=0} = \frac{\partial}{\partial x_1^i}\frac{\partial}{\partial x_2^i} v_0^\alpha(x), \\
&\cdots\cdots\cdots\cdots\cdots\cdots\cdots\cdots\cdots\cdots
\end{aligned}
$$

respectively. The integrability of (16.1), (16.2) implies the integrability of the Lie-Bäcklund equations (16.6), (16.7). Let the solutions of (16.6) and (16.7) be given by

$$
\begin{aligned}
&t'=\tau,\\
&x'=x,\\
&u'^{\alpha}=F_1^{\alpha}\Big(x,u_0,\underset{1}{u}_0,\cdots;\tau\Big),\\
&u_i'^{\alpha}=D_iF_1^{\alpha}\Big(x,u_0,\underset{1}{u}_0,\cdots;\tau\Big),\\
&u_{i_1i_2}'^{\alpha}=D_{i_1}D_{i_2}F_1^{\alpha}\Big(x,u_0,\underset{1}{u}_0,\cdots;\tau\Big),\\
&\cdots\cdots\cdots\cdots\cdots\cdots\cdots\cdots\cdots
\end{aligned}
\tag{16.8}
$$

with $F_1^{\alpha}(x,u_0,\underset{1}{u}_0,\cdots;0)=u_0^{\alpha}(x)$, and

$$
\begin{aligned}
&t'=\tau,\\
&x'=x,\\
&v'^{\alpha}=F_2^{\alpha}(x,v_0,\underset{1}{u}_0,\cdots;\tau),\\
&v_i'^{\alpha}=D_iF_2^{\alpha}(x,v_0,\underset{1}{v}_0,\cdots;\tau),\\
&v_{i_1i_2}'^{\alpha}=D_{i_1}D_{i_2}F_2^{\alpha}(x,v_0,\underset{1}{v}_0,\cdots;\tau),\\
&\cdots\cdots\cdots\cdots\cdots\cdots\cdots\cdots\cdots
\end{aligned}
\tag{16.9}
$$

with $F_2^{\alpha}(x,v_0,\underset{1}{v}_0,\cdots;0)=v_0^{\alpha}(x)$, respectively. The functions $F_i^{\alpha}(i=1,2)$ appearing in (16.8), (16.9) follow directly from the functions $K_i^{\alpha}(i=1,2)$ via Taylor expansions (e.g., see Examples 3 and 4 in §12). The group properties of equations (16.8), (16.9) allow us to establish a one-to-one correspondence between these Lie-Bäcklund transformation groups. This correspondence is given by

$$
\begin{aligned}
&u^{\alpha}=F_1^{\alpha}(x,F_2(x,v,\underset{1}{v},\cdots;-t),DF_2(x,v,\underset{1}{v},\cdots;-t),\cdots;t),\\
&u_i^{\alpha}=D_iF_1^{\alpha}(x,F_2(x,v,\underset{1}{v},\cdots;-t),DF_2(x,v,\underset{1}{v},\cdots;-t),\cdots;t),\\
&\cdots\cdots\cdots\cdots\cdots\cdots\cdots\cdots\cdots\cdots\cdots\cdots
\end{aligned}
\tag{16.10}
$$

with inverse

$$
\begin{aligned}
&v^{\alpha}=F_2^{\alpha}(x,F_1(x,u,\underset{1}{u},\cdots;-t),DF_1(x,u,\underset{1}{u},\cdots;-t),\cdots;t),\\
&v_i^{\alpha}=D_iF_2^{\alpha}(x,F_1(x,u,\underset{1}{u},\cdots;-t),DF_1(x,u,\underset{1}{u},\cdots;-t),\cdots;t),\\
&\cdots\cdots\cdots\cdots\cdots\cdots\cdots\cdots\cdots\cdots\cdots\cdots
\end{aligned}
\tag{16.11}
$$

for each t and x. In this interpretation, the groups of transformations of the form (16.5) represent the projection on the space $\mathcal{F}_{x,\ t}$ of the action of the Lie-Bäcklund groups defined by (16.8), (16.9).

The one-to-one correspondence given by (16.10), (16.11) is important if we suppose that the system described by the evolution equation (16.1) admits a point transformation group G_1 which can be a group of Lie point transformations, Lie tangent transformations, or Lie-Bäcklund tangent transformations:

$$(16.12)\qquad G_1:\begin{cases} t'=\tau\left(t, x, u, \underset{1}{u}, \cdots; a\right), \\ x'=f\left(t, x, u, \underset{1}{u}, \cdots; a\right), \\ u'=\phi\left(t, x, u, \underset{1}{u}, \cdots; a\right), \\ \underset{1}{u}'=\underset{1}{\psi}\left(t, x, u, \underset{1}{u}, \cdots; a\right), \\ \cdots\cdots\cdots\cdots\cdots\cdots \end{cases}$$

The correspondence (16.10), (16.11) converts the invariance group G_1 into the isomorphic group G_2 which is admitted by the evolution equation (16.2):

$$\begin{aligned} &t'=\tau\,(t, z; a)|_{(16.10)} \overset{\text{Def}}{=} \tilde{\tau}\left(t, x, v, \underset{1}{v}, \cdots; a\right), \\ &x'=f\,(t, z; a)|_{(16.10)} \overset{\text{Def}}{=} \tilde{f}\left(t, x, v, \underset{1}{v}, \cdots; a\right), \\ (16.13)\qquad G_2:\quad &v'=F_2(x'|_{(16.12)}, F_1(z'; -t')|_{(16.12)}, \\ &\qquad DF_1(z'; -t')|_{(16.12)}, \cdots; t'|_{(16.12)})|_{(16.10)} \\ &\overset{\text{Def}}{=} \tilde{\phi}\left(t, x, v, \underset{1}{v}, \cdots; a\right), \end{aligned}$$

where $z'=(x', u', \underset{1}{u}', \cdots)$ and z is defined similarly without primes. This is the analogue of (15.19), and similarly G_2 is in general a Lie-Bäcklund transformation group, although G_1 may be a group of Lie point transformations or Lie tangent transformations. If (16.12), (16.13) are admitted by (16.1), (16.2), then they are also admitted by (16.6), (16.7), respectively.

The preceding transformations (16.10), (16.11), and (16.13) apply independently of the linearity of the equations (16.6), (16.7), but in the remainder of this section we shall confine the discussion to the application of these results to linear equations describing quantum-mechanical systems. Here these systems will be described exclusively in the coordinate representation of the Schrödinger picture (e.g., Dirac [1]) and for brevity will be called Schrödinger systems. It follows directly as an application of the preceding development that every quantum-mechanical Schrödinger group is equivalent to a Lie-Bäcklund transformation group. The latter is defined in general only on the manifold of solutions of the corresponding time-dependent Schrödinger equation for the initial-value problem

$$i\hbar\frac{\partial u}{\partial t}=\hat{H}u, \qquad u|_{t=0}=u_0(x), \tag{16.14}$$

where H is a self-adjoint operator of the form

$$\hat{H}=-\frac{\hbar^2}{2m}\Delta+V(x), \tag{16.15}$$

and where $\hbar$ is Planck's constant divided by 2π, m is the mass of the particle, Δ is the three-dimensional Laplacian, and $x\in\mathbb{R}^3$.

The unitary representation $\{e^{-i\hat{H}t/\hbar}\}_{t\in\mathbb{R}}$ generated by the self-adjoint Hamiltonian operator $\hat{H}$ of the form (16.15) is equivalent to the induced one-parameter Lie-Bäcklund transformation group corresponding to the Lie-Bäcklund operator

$$X=\eta\frac{\partial}{\partial u}+(D_i\eta)\frac{\partial}{\partial u_i}+(D_{i_1}D_{i_2}\eta)\frac{\partial}{\partial u_{i_1i_2}}+\cdots, \tag{16.16}$$

defined on the solutions of (16.14), where

$$\eta=i\frac{\hbar}{2m}(u_{11}+u_{22}+u_{33})-\frac{i}{\hbar}V(x)u. \tag{16.17}$$

This statement is immediate with the identification of (16.14) where $\hat{H}$ is given by (16.15) with (16.1), $e^{-i\hat{H}t/\hbar}u_0(x)$ with $K_1(u_0(x);\, t)$, and the Lie-Bäcklund operator (16.16), (16.17) with the Lie-Bäcklund equation (16.6).

It follows from the preceding that the one-to-one correspondence (16.10), (16.11) is applicable to any two Schrödinger systems. This latter statement is a translation of a well-known quantum-mechanical map symbolically represented by $v_t(x)=e^{-iH_2t/\hbar}e^{i\hat{H}_1t/\hbar}u_t(x)$, the Hamiltonians $\hat{H}_1$, $\hat{H}_2$ being identified with (16.1), (16.2), respectively.

Proceeding, we now establish the fact that every quantum-mechanical constant of the motion of a Schrödinger system (16.14) corresponds to a Lie-Bäcklund operator on the manifold of solutions of (16.14). In order to establish this correspondence, we first review the definition of a quantum-mechanical constant of the motion and then trace the well-known result that the self-adjoint operator $\hat{Q}$ representing a quantum-mechanical constant of the motion is also an invariance operator admitted by (16.14).

A self-adjoint operator $\hat{Q}$ represents a constant of the motion of a Schrödinger system (16.14) described by a Hamiltonian (16.15) if

$$\frac{d}{dt}\int_{\mathbb{R}^3}\overline{u_t(x)}\;(\hat{Q}u_t(x))\,d^3x=0 \tag{16.18}$$

for arbitrary square-integrable u satisfying (16.14), where $\overline{u_t(x)}$ is the complex conjugate of $u_t(x)$. Equation (16.18) implies by differentiation under the integral sign that

$$\left(i\hbar\frac{\partial\hat{Q}}{\partial t}+[\hat{Q},\hat{H}]\right)u|_{(16.14)}\equiv 0, \tag{16.19}$$

where $[\hat{Q},\hat{H}]=\hat{Q}\hat{H}-\hat{H}\hat{Q}$ is the commutator of the operators $\hat{Q}$, $\hat{H}$. It follows directly from (16.19) by simple rearrangement that

$$\hat{Q}\left(i\hbar\frac{\partial}{\partial t}-\hat{H}\right)u|_{(16.14)}\equiv 0, \tag{16.20}$$

if and only if

$$\left(i\hbar\frac{\partial}{\partial t}-\hat{H}\right)\hat{Q}u|_{(16.14)}\equiv 0, \tag{16.21}$$

i.e., $\hat{Q}$ is an invariance operator admitted by (16.14) if $\hat{Q}$ acts invariantly on the manifold of solutions of (16.14). These results lead directly to the following statement.

THEOREM 16.1. *Every linear self-adjoint operator*

$$\hat{Q}=q(x,t)+q^{i}(x,t)\frac{\partial}{\partial x^{i}}+q^{i_1i_2}(x,t)\frac{\partial^2}{\partial x^{i_1}\partial x^{i_2}}+\cdots, \tag{16.22}$$

representing a constant of the motion of a quantum-mechanical system (16.14) *described by a given Hamiltonian operator of the form* (16.15), *is equivalent to a Lie-Bäcklund invariance operator*

$$X=\eta\frac{\partial}{\partial u}+(D_i\eta)\frac{\partial}{\partial u_i}+(D_{i_1}D_{i_2}\eta)\frac{\partial}{\partial u_{i_1i_2}}+\cdots, \tag{16.23}$$

where

$$\eta=\hat{Q}u=q(x,t)u+q^{i}(x,t)u_i+q^{i_1i_2}(x,t)u_{i_1i_2}+\cdots, \tag{16.24}$$

on the solutions of (16.14).

[Note that $(i\hbar\partial/\partial t)^k\equiv(\hat{H})^k$ arbitrary positive k on the solutions of (16.14).]

It is now possible to illustrate all the features of this section by simply taking for the system (16.1) the quantal free particle and for (16.2) the quantal particle in a uniform external field. Specifically, the quantum-mechanical free particle in one spacial dimension is described by the time-dependent Schrödinger equation

$$i\hbar\frac{\partial u}{\partial t}=\frac{-\hbar^2}{2m_0}\frac{\partial^2}{\partial x^2}u \tag{16.25}$$

with the integral representation of the solution of the initial-value problem (16.1) given by

$$u_t(x)=\int_{-\infty}^{\infty}db\left(\frac{m_0}{2\pi i\hbar t}\right)^{1/2}\exp\left[-\frac{m_0}{2\pi\hbar t}(x-b)^2\right]u_0(b), \tag{16.26}$$

where $u_0\in\mathfrak{L}^2(\mathbb{R})$. It follows via a Taylor expansion that the equivalent one-parameter Lie-Bäcklund transformation group is given by (16.8) with

$$F_1\left(x, u, \underset{1}{u}, \cdots; \tau\right) = u + \sum_{k=1}^{\infty} \frac{(i\hbar\tau/2m_0)^k}{k!}(D_x^{2k}u), \tag{16.27}$$

and corresponding Lie-Bäcklund operator defined on the manifold of solutions of (16.25) is given by (16.16) with

$$\eta = \frac{i\hbar}{2m_0} u_{xx}. \tag{16.28}$$

The second system, the quantal particle in a uniform external field of strength k in one spatial dimension, is described in Cartesian coordinates by the time-dependent Schrödinger equation

$$i\hbar\frac{\partial v}{\partial t} = \frac{-\hbar^2}{2m}\frac{\partial^2}{\partial x^2}v + kxv, \tag{16.29}$$

with the integral representation of the solution of the initial-value problem (16.14) given by

$$v_t(x) = \int_{-\infty}^{\infty} db \left(\frac{m}{2\pi i\hbar t}\right)^{1/2} \exp\left[-\frac{m(x-b)^2}{2i\hbar t} - \frac{ikxt}{\hbar} - \frac{ik^2t^3}{3!\hbar}\right] v_0(x), \tag{16.30}$$

where $v_0 \in \mathcal{L}^2(\mathbb{R})$. This implies via a Taylor expansion that the equivalent one-parameter Lie-Bäcklund transformation group is given by (16.9) with

$$\begin{aligned} &F_2\left(x, v, \underset{1}{v}, \cdots, \tau\right) \\ &= \exp\left[\frac{i\tau^3k^2}{6\hbar m} - \frac{ikx\tau}{\hbar}\right] \sum_r \sum_s \frac{(i\hbar\tau/2m)^r}{r!}\frac{(k\tau^2/2m)^s}{s!}(D_x^{2r+s}v), \end{aligned} \tag{16.31}$$

and the corresponding Lie-Bäcklund operator defined on the manifold of solutions of (16.29) is given by (16.16) with

$$\eta = \frac{i\hbar}{2m}v_{xx} - \frac{i}{\hbar}kxv. \tag{16.32}$$

Therefore, with the identification $v_0(x) = u_0(x)$, equations (16.10), (16.11) provide a one-to-one correspondence between the Lie-Bäcklund transformations groups (16.8), (16.9), where F_1, F_2 are given by (16.27), (16.31), respectively.

The self-adjoint operator

$$\hat{Q} = \frac{\hbar}{i}\frac{\partial}{\partial x} \tag{16.33}$$

is a constant of the motion of the system (16.25), and according to Theorem 16.1

it is equivalent to the Lie operator

$$X=\frac{\hbar}{i}u_x\frac{\partial}{\partial u}. \tag{16.34}$$

The operator (16.34) generates the Lie transformation group

$$\begin{aligned} t'&=t,\\ x'&=x,\\ u'&=\sum_{n=0}^{\infty}\frac{(-i\hbar a)^n}{n!}(D_x^n u), \end{aligned} \tag{16.35}$$

with group parameter a. Therefore, if we identify the infinite extension of the one-parameter group of transformations (16.35) with (16.12), we obtain via (16.10) [where F_1, F_2 are given by (16.27), (16.31), respectively] an isomorphic group of transformations (16.13) admitted by (16.29).

This automatically establishes a one-to-one correspondence between the Lie operator (16.34) and its induced image for (16.29). This also establishes a one-to-one correspondence between the constant of the motion (16.33) and the constant of the motion for (16.29) that is equivalent (Theorem 16.1) to the induced image of the Lie operator (16.34). In this latter regard, the operator formulation of maps (16.10), (16.11) has been previously employed to establish this one-to-one correspondence between the constants of the motion of two Schrödinger systems with the same spacial dimensionality (Anderson, Shibuya, and Wulfman [1]). It is illustrated there with an example for the free particle and harmonic oscillator in one spatial dimension.

More generally, Lie [7] (p. 357) established that the equation

$$\frac{\partial u}{\partial t}=-\frac{\partial^2 u}{\partial x^2}, \tag{16.36}$$

admits the following Lie operators:

$$\begin{aligned} &X_1=\frac{\partial}{\partial x}, \qquad X_2=\frac{\partial}{\partial t},\\ &X_3=x\frac{\partial}{\partial x}+2t\frac{\partial}{\partial t}, \qquad X_4=2t\frac{\partial}{\partial x}+xu\frac{\partial}{\partial u},\\ &X_5=xt\frac{\partial}{\partial x}+t^2\frac{\partial}{\partial t}+\left(\tfrac{1}{4}x^2u-\tfrac{1}{2}tu\right)\frac{\partial}{\partial u}. \end{aligned} \tag{16.37}$$

The set (16.37) (e.g., under the change of variables $t\to-2it/\hbar$, $x\to x/\hbar$ and overall multiplication by $\pm i\hbar$) yields, in addition to $\hat{Q}_1$ given by (16.33), four other constants of the motion of (16.25). Treatments similar to that given for $\hat{Q}_1$ are possible for the other constants of the motion. In fact, the treatment of all these operators is subsumed in the following discussion of the three-dimensional analogues of the systems (16.25), (16.29).

The time-dependent Schrödinger equation for the three-dimensional quantal free particle

$$(16.38)\qquad i\frac{\partial}{\partial t}u=-\frac{1}{2m_0}\left(\left[\frac{\partial}{\partial x^1}\right]^2+\left[\frac{\partial}{\partial x^2}\right]^2+\left[\frac{\partial}{\partial x^3}\right]^2\right)u$$

admits a 12-parameter group of Lie point transformations (Niederer [1], Barut [1], Barut and Rączka [1]) given by

$$(16.39)\qquad \begin{aligned} t'&=d^2\frac{t+b}{1+\alpha\,(t+b)},\\ x'&=d\frac{Rx+Vt+a}{1+\alpha\,(t+b)},\\ u'&=f_g(t,x)u, \end{aligned}$$

with

$$\begin{aligned} f_g(t,x)&=C_g\left[U_g(t)\right]^{3/2}\exp\left[\frac{-im_0}{2U_g(t)}k_g(t,x)\right],\\ U_g(t)&=1+\alpha\,(t+b),\\ k_g(t,x)&=\alpha x^ix^i+2\,(Rx)^i\left[\alpha a^i-(1+\alpha b)V^i\right]+\alpha a^ia^i\\ &\quad-(1+\alpha b)tV^iV^i+2\alpha ta^iV^i,\\ C_g&=d^{-3/2}, \end{aligned}$$

where $d\in\mathbb{R}-0$, $b\in\mathbb{R}$, $a=(a^1,a^2,a^3)\in\mathbb{R}^3$, $V=(V^1,V^2,V^3)\in\mathbb{R}^3$, R belongs to the 3×3 real matrix representation of the special orthogonal group in three dimensions, $\alpha\in\mathbb{R}$, and the dummy indices are summed from 1 to 3. In (16.38), (16.39) and in the remainder of this discussion, the physical constant $\hbar$ has been set equal to 1. Therefore, if we identify the infinite extension of the group of transformations (16.39) with (16.12), we obtain via (16.10), (16.11), where

$$(16.40)\qquad \begin{aligned} F_1&=\sum_{n_1}^{\infty}\sum_{n_2}^{\infty}\sum_{n_3}^{\infty}\frac{(i\tau/2m_0)^{n_1+n_2+n_3}}{n_1!n_2!n_3!}\\ &\quad\times\left(D_{x^1}^{2n_1}D_{x^2}^{2n_2}D_{n^3}^{2n_3}u\right), \end{aligned}$$

$$(16.41)\qquad \begin{aligned} F_2&=e^{[(i\tau^3k^2/6m)-ikx\tau]}\sum_r\sum_s\sum_{n_2}\sum_{n_3}\\ &\quad\times\frac{(i\tau/2m)^r}{r!}\frac{(k\tau^2/2m)^s}{s!}\left(D_{x^1}^{2r+s}D_{x^2}^{2n_2}D_{x^3}^{2n_3}v\right), \end{aligned}$$

an isomorphic group of transformations (16.13) admitted by the three-dimensional quantal particle in a uniform external field of strength k parallel to the x^1-direction.

We conclude with the observation that in this way the group of transformations (16.39) induces a common Lie-Bäcklund transformation group structure for all three-dimensional Schrödinger systems.

Chapter 4

Some Applications of Bäcklund Transformations

A number of nonlinear evolution equations, such as the Burgers, sine-Gordon, nonlinear Schrödinger, and Korteweg–de Vries (KdV) equations, are known to share some remarkable properties. Some understanding of their similarities has been due to the discovery that these equations are members of a class that can be solved by an inverse procedure.[1] Typically they possess stable steady-state solutions with remarkable properties such as stability through interactions and (nonlinear) superposability.

While a uniform approach, which can be applied systematically to all such equations, is not at hand, it appears that the transformation and invariance properties of the equations may be the key principles. The Burgers equation is directly transformable to the linear diffusion equation. The sine–Gordon equation, which arose many years ago in connection with a transformation problem in differential geometry, is known to possess a Bäcklund transformation. From this transformation many of the interesting properties of the sine–Gordon equation can be derived.

In this chapter we describe some applications of Bäcklund transformations to a variety of physical problems, including nonlinear optics, lattices, diffusion, and nonlinear waves. Since our central interest here is the physical implication of the transformations, they are not derived but just stated.

§17. Nonlinear Optics[2–4]

Recent advances in laser technology have led to the production of coherent optical pulses having durations in the picosecond (10^{-12} sec) regime—so-called *ultrashort optical pulses*. The resonant interaction of radiation and matter on

[1]See e.g. M. J. Ablowitz, D. J. Kaup, A. C. Newell, and H. Segur, *The initial value solutions for the sine–Gordon equation*, Phys. Rev. Lett., 31 (1973), p. 125.

[2]G. L. Lamb, Jr. *Propagation of ultra-short optical pulses*, in Honor of Philip M. Morse, H. Feshbach and K. U. Ingard, ed., M.I.T. Press, Cambridge, Mass., 1969, p. 88.

[3]G. L. Lamb, Jr. *Analytical descriptions of ultra-short optical pulse propagation in a resonant medium*, Rev. Modern Phys., 43 (1971), p. 99.

[4]A. Barone, F. Esposito, C. J. Magee, and A. C. Scott, *Theory and application of the sine–Gordon equation*, Riv. Nuovo Cimento (2), 1 (1971), p. 227.

such a short time scale generates phenomena which, as a result of the quantum-mechanical coherence effects, cannot be described by the rate-equation analysis developed for the treatment of long pulses.

Under the assumptions of vanishing bandwidth and neglect of nonresonant losses, Lamb[5] shows that the sine–Gordon equation

$$\frac{\partial^2\sigma}{\partial\eta\,\partial\nu}=\sin\sigma \tag{17.1}$$

is the fundamental equation for the calculation of the electric field and polarization of the medium.

Given a solution σ_0 of (17.1), the kth solution is computed from the (k-1)st through the Bäcklund transformation (see Chapter 1, §7)

$$\frac{\partial}{\partial\eta}\left(\frac{\sigma_k-\sigma_{k-1}}{2}\right)=a_k\sin\tfrac{1}{2}(\sigma_k+\sigma_{k-1}), \tag{17.2a}$$

$$\frac{\partial}{\partial\nu}\left(\frac{\sigma_k+\sigma_{k-1}}{2}\right)=a_k^{-1}\sin\tfrac{1}{2}(\sigma_k-\sigma_{k-1}). \tag{17.2b}$$

For $\sigma_0=0$, equations (17.2) generate the solution

$$\sigma_1=4\tan^{-1}[\exp(a_1\eta+\nu/a_1)], \tag{17.3}$$

which becomes

$$\phi_1(x,t)=4\tan^{-1}\left[\exp\left[\frac{x-u_1t}{\sqrt{1-u_1^2}}\right]\right] \tag{17.4}$$

under the transformations

$$\eta=\tfrac{1}{2}(x+t),\quad \nu=\tfrac{1}{2}(x-t),\quad u=\frac{1-a^2}{1+a^2}. \tag{17.5}$$

In these physical variables the Bäcklund transformation has created an evenly spaced array of "kinks," or regions where ϕ changes by 2π, which propagates uniformly for all x and all t. Equation (17.4) represents the propagation of a single kink (one soliton).

As indicated in Chapter 1, §7, if we start with a known solution σ_0, generate σ_1 by means of (17.2) through a_1, and generate σ_2 through a_2, then there is a σ_3 which is generated from σ_1 through a_2 and also from σ_2 through a_1, and these solutions are related through the nonlinear superposition

$$\tan\left(\frac{\sigma_3-\sigma_0}{4}\right)=\frac{a_1+a_2}{a_1-a_2}\tan\left(\frac{\sigma_1-\sigma_2}{4}\right). \tag{17.6}$$

[5]Lamb, op. cit.

From (17.5), the same relations must be satisfied by solutions of

(17.7) $$\phi_{xx}-\phi_{tt}=\sin\phi.$$

With $\phi_0=0$, ϕ_1 as in (17.4), and

(17.8) $$\phi_2=4\tan^{-1}\left[\exp\left(-\frac{x-u_2t}{\sqrt{1-u_2^2}}\right)\right],$$

where

$$\frac{a_1+a_2}{a_1-a_2}=\frac{\sqrt{\dfrac{1-u_1}{1+u_1}}-\sqrt{\dfrac{1-u_2}{1+u_2}}}{\sqrt{\dfrac{1-u_1}{1+u_1}}+\sqrt{\dfrac{1-u_2}{1+u_2}}}=k(u_1,u_2),$$

a solution of (17.7) obtained from (17.6) is

(17.9) $$\tan\left(\frac{\phi_3}{4}\right)=k(u_1,u_2)\frac{\exp\left[\dfrac{x-u_1t}{\sqrt{1-u_1^2}}\right]-\exp\left[-\dfrac{x-u_2t}{\sqrt{1-u_2^2}}\right]}{1+\exp\left[\dfrac{x-u_1t}{\sqrt{1-u_1^2}}\right]\exp\left[-\dfrac{(x-u_2t)}{\sqrt{1-u_2^2}}\right]}.$$

This is a "two-kink" (two-soliton) solution, since for $u_1<u_2$ the total change in ϕ_3 is equal to 4π. Using the Lorentz transformation

$$x\to x^1=\frac{x-ut}{\sqrt{1-u^2}},$$

$$t\to t^1=\frac{t-ux}{\sqrt{1-u^2}}$$

to put the problem into a center-of-mass coordinate system (equivalently, choosing $u_2=-u_1$), we obtain the kink-kink collision

$$\phi=4\tan^{-1}\left[\frac{u\sinh\left(x/\sqrt{1-u^2}\right)}{\cosh\left(ut/\sqrt{1-u^2}\right)}\right].$$

If we set $u_1=u_2$, we find $k\equiv0$, thereby proving that only the traveling-wave solutions (waves of permanent profile) contain either one or an infinite number of kinks.

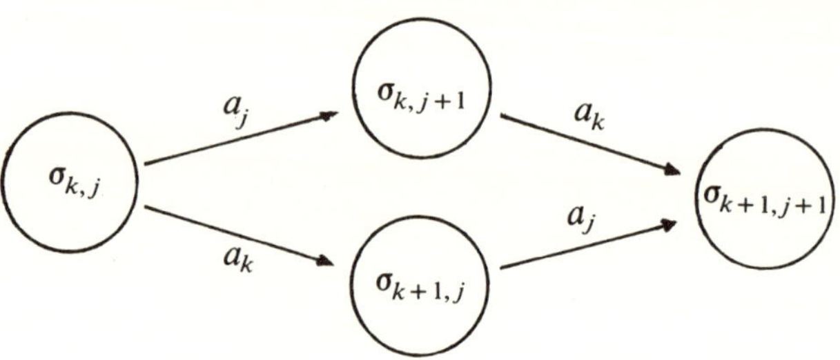

FIG. 10. Diagram for equation (17.10).

By a modification of (17.6) due to McLaughlin and Scott[6] four solutions of (17.1) can be related in terms of two constants a_k and a_j such that

$$\left.\begin{aligned}\sigma_{k,j+1}&=\beta_{a_j}\sigma_{k,j}\\ \sigma_{k+1,j}&=\beta_{a_k}\sigma_{k,j}\\ \phi_{k+1,j+1}&=\beta_{a_j}\beta_{a_k}\sigma_{k,j}=\beta_{a_k}\beta_{a_j}\sigma_{k,j}\end{aligned}\right\}k,j=1,2,\cdots, \tag{17.10}$$

where β_{a_j} is the Bäcklund transformation operator associated with a_j [equation (17.2)]. Using a diagram like that employed by Bianchi, this association is shown in Fig. 10.

Barnard[7] shows (Fig. 11) that through a series of these Bäcklund transformations one can generate the N-soliton solutions of the sine–Gordon equation. The number in parentheses represents the number of solitons. The analytic expressions for these solutions are

$$\begin{aligned}&\sigma_{j+1,j}=0,\\ &\quad\sigma_{j,j}=4\tan^{-1}\left[e^{-K_jx+(1/K_j)t+\gamma_{jj}}\right],\\ &\quad\gamma_{jj}=\text{const}\quad\text{and}\quad(-1)^j/K_j<0\qquad\text{if } j>0.\end{aligned}$$

If $j>k$, then

$$\sigma_{k,j}=\sigma_{k+1,j-1}+4\tan^{-1}\left[\frac{\dfrac{1}{K_j}-\dfrac{1}{K_k}}{\dfrac{1}{K_k}+\dfrac{1}{K_j}}\tan\left(\frac{\sigma_{k,j-1}-\sigma_{k+1,j}}{4}\right)\right],$$

$$(-1)^k/K_k<(-1)^j/K_j.$$

---

[6]D. W. McLaughlin and A. C. Scott, *A restricted Bäcklund transformation*, submitted.

[7]T. Barnard, *2Nπ ultrashort pulses*, Phys. Rev. A, 7 (1973), p. 373.

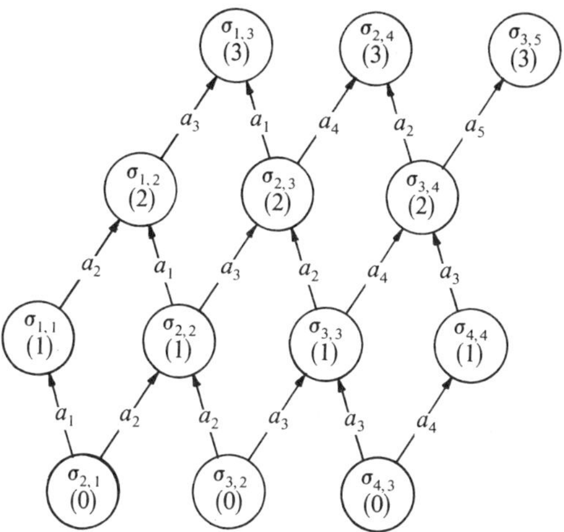

FIG. 11. Diagram for N soliton solutions of the sine–Gordon equation.

To generate an N-soliton solution it is necessary to develop the expressions for all soliton solutions of order less than N. Caudrey et al.[8] claim to have obtained an N-soliton formula for the sine–Gordon equation by direct computation. The key to their work is the transformation.

$$\sigma=\cos^{-1}\left[1+2\left(\frac{\partial^2}{\partial t^2}-\frac{\partial^2}{\partial x^2}\right)\ln f(x,t)\right],$$

where $f(x,t)=\det|M|$ and M is the $N\times N$ matrix

$$M_{kj}=\frac{(a_k a_j)^{1/2}}{a_k+a_j}\left[e^{K_k x+\beta_k t+\gamma_k}+(-1)^{k+j}e^{-K_j x-\beta_j t+\gamma_j}\right],$$

with

$$\gamma_j=\text{const},\qquad a_j=\text{const},$$
$$K_j=a_j+a_j^{-1},\qquad \beta_j=a_j-a_j^{-1}.$$

§18. Solitons and the KdV Equation[9]

The scaled Korteweg–de Vries equation in conservation form is

$$u_t+\left(6u^2+u_{xx}\right)_x=0. \tag{18.1}$$

When a potential function w is introducted by setting $u=-w_x$, it follows from

[8]P. J. Caudrey, J. D. Gibbon, J. C. Eilbeck and R. K. Bullough, *Exact multisoliton solutions of the self induced transparency and sine–Gordon equations*, Phys. Rev. Lett. 30 (1973), p. 237–239.

[9]H. D. Wahlquist and F. B. Estabrook, *Bäcklund transformation for solutions of the Korteweg–de Vries equation*, Phys. Rev. Lett., 31 (1973), p. 1386.

(18.1) that w satisfies the equation

$$w_t = 6w_x^2 - w_{xxx}. \tag{18.2}$$

In terms of w the single-soliton solutions are obtainable from the completely integrable Pfaffian system

$$w_x = w^2 - k^2 = -u, \tag{18.3a}$$

$$w_t = -4k^2(w^2 - k^2) = 4k^2u, \tag{18.3b}$$

k an arbitrary constant, which is shown to satisfy (18.2) by direct substitution.[10] With $\xi = -k(x - 4k^2t) + c$ the regular integral of (18.3) is

$$w = k\tanh\xi, \qquad u = -w_x = k^2\operatorname{sech}^2\xi. \tag{18.4}$$

The arbitrary parameter k determines both the amplitude of the soliton, $A = k^2$, and its speed $V = 4k^2$, while the integration constant c specifies the initial position or phase of the soliton.

Since the system in (18.3) is invariant under the transformation $w \to k^2/w$, it also has the singular solution

$$w_1 = k\coth\xi, \qquad u_1 = -k^2\operatorname{csch}^2\xi. \tag{18.5}$$

Both the regular and the singular solution have the same asymptotic values.

Let w be any solution of (18.2), and $u = -w_x$ be the associated solution of (18.1). A different solution (w^1, u^1) is then defined by the already generated (§7) Bäcklund transformation[11]

$$\begin{aligned} w_x^1 + w_x &= (w^1 - w)^2 - k^2, \\ w_t^1 + w_t &= 4\left[k^2u^1 + u^2 - u(w^1 - w)^2 - u_x(w^1 - w)\right], \end{aligned} \tag{18.6}$$

where k is an arbitrary parameter. By direct calculation it is easily seen that (18.6) is integrable and that w^1 satisfies (18.2). Since integrability of (18.6) is assured, only the first equation of (18.6) will henceforth be considered, in the form

$$(u^1 + u) = k^2 - (w^1 - w)^2. \tag{18.7}$$

Starting from an arbitrary solution w, a sequence of solutions will now be generated using (18.7). First $w_1 = w_1(k_1)$ is calculated with

$$u_1 + u = k_1^2 - (w_1 - w)^2,$$

then $w_2 = w_2(k_2)$ with

$$u_2 + u = k_2^2 - (w_2 - w)^2,$$

[10] See also H. D. Wahlquist and F. B. Estabrook, *Prolongation structures of nonlinear evolution equations*, J. Mathematical Phys., 16 (1975), p. 1.

[11] Equation (18.6) differs slightly from (7.55) because of the 6 in (18.2).

and then $w_{12}=w_{12}(k_1, k_2)=w_{21}(k_2, k_1)$ with the nonlinear superposition (cf. §7)

$$w_{12}=w+\frac{k_2^2-k_1^2}{w_2-w_1}. \tag{18.8}$$

In the previous remarks the subscripts serve as a shorthand notation for the indicated parametric dependence. Equation (18.8) expresses the second-order transformation of any solution as the nonlinear superposition of two first-order transformations of that solution.

Equation (18.2) admits the trivial solution, so a single soliton can be generated, via (18.7), from the "soliton vacuum." When the single-soliton solution is used in (18.7), a 2-soliton solution is generated. When applied recursively, each successive transformation adds one new soliton to the previous solution. Thus, after n applications the known "pure" n-soliton solution is generated.

However, we need not begin with the "vacuum solution." A similar sequence (ladder) of solutions to the KdV equation can be constructed by recursive application of (18.7) to *any starting solution*. Each step up the ladder does not require the integration of a new more complicated system. The actual integration of (18.7) is required *only* for the first step up the ladder from any starting solution. Succeeding steps are reached algebraically by employing the nonlinear superposition (18.8).

In (18.8) the starting solution was arbitrary, so by induction the solution at the nth step of the ladder is given by

$$w_{(n)}=w_{(n-2)}+\frac{k_n^2-k_{n-1}^2}{w_{(n-1)'}-w_{(n-1)}}, \qquad n>1, \tag{18.9}$$

where the subscript (n) denotes the set of n parameters $\{k_1,\cdots,k_n\}$, $(n)'$ denotes $\{k_1, k_2,\cdots,k_{n+1}\}$, and $w_{(0)}=w$.

The third-order $(n=3)$ transformed solution is

$$\begin{aligned} w_{(3)}=w_{123}&=w_{(0)}+\frac{k_3^2-k_2^2}{w_{(2)'}-w_{(2)}} \\ &=w+\frac{k_3^2-k_2^2}{w_{13}-w_{12}}. \end{aligned} \tag{18.10}$$

Using (18.9) again, with $n=2$, the denominator of (18.10) can be expressed entirely in terms of first-order solutions, whereupon

$$w_{123}=\frac{k_1^2w_1(w_2-w_3)+k_2^2w_2(w_3-w_1)+k_3^2w_3(w_1-w_2)}{k_1^2(w_2-w_3)+k_2^2(w_3-w_1)+k_3^2(w_1-w_2)}. \tag{18.11}$$

A permutation symmetry with respect to (k, w) pairs is evident.

The hierarchy of pure multisoliton solutions is given by the previous results when the starting solution is chosen to be the vacuum solution $w=w_{(0)}=0$. Each first-order function w_i is then a single-soliton solution corresponding to the parameter k_i, as given by (18.4) in the regular case or (18.5) in the singular one. Since any set of parameters may be used, and any combination of regular or singular w_i, it follows that a large family of different hierarchies of solutions can be obtained. For any given parameter set, however, only one choice of the w_i will produce regular solutions at every step of the ladder. The *required choice* is not immediately obvious. It is not true that all w_i must be regular.

Let us choose the regular soliton w_1 of (18.4) for the first step, and using (18.7) transform to the two-soliton solution w_{12}. Writing that result in accordance with (18.8), the only regular solution obtainable is given by

$$w_{12}=\frac{k_2^2-k_1^2}{w_2{}^*-w_1}, \qquad k_2>k_1, \tag{18.12}$$

where the asterisk denotes the singular solution. This is the *only* case in which the denominator never vanishes for any values of ξ_1 and ξ_2. To maintain regularity successive transformations with (18.7) must be applied with monotonically increasing parameters, i.e., $k_1<k_2<k_3<\cdots$.

§19. Constants of the Motion and Conservation Laws[12, 13]

Bäcklund transformations provide a rational means for finding eigenvalue problems (through the relationship of the Riccati equation to a linear second-order equation) to associate with certain evolution equations. One of the other valuable uses which these transformations can serve is the generation of *conservation laws* and *motion invariants* (constants of motion). While definitions and examples of these concepts have been specified in §14, we need our own peculiar twist here. Therefore we discuss the two ideas briefly.

Let L denote a linear space and N a (nonlinear) operator mapping L into L, and for $0\leqq t<\infty$, let $\phi(t)$ denote an element of L. We shall examine examples of the abstract evolution equation

$$\phi_t=M(\phi) \tag{19.1}$$

in L. Let $I[\cdot]$ be a functional mapping L into the complex numbers. If

$$\frac{d}{dt}I[\phi]=0 \tag{19.2}$$

for *all* solutions ϕ of (19.1), the functional I is said to be a *constant of the motion* (motion invariant).

In practice these constants of the motion arise in most physical situations through the concept of a *conservation law*. For a more precise definition we

[12] M. J. Ablowitz, D. J. Kaup, A. C. Newell, and H. Segur, *The inverse scattering transform—Fourier analysis for nonlinear problems*, Phys. Rev. Lett., 31 (1973), p. 125.

[13] A. C. Scott, F. Y. F. Chu, and D. W. McLaughlin, *The soliton, a new concept in applied science*, Proc. IEEE, 61 (1973), p. 1443.

restrict our attention to the case of partial differential equations, with L a set of real-valued functions of the continuous real variable x, $-\infty<x<\infty$. Let $D[\cdot]$ and $F[\cdot]$ be a pair of (nonlinear) operators mapping L into L such that $D[\phi(x,t)]$ and $F[\phi(x,t)]$ depend locally upon x and t. If

$$\frac{\partial}{\partial t} D\left[\phi(x,t)\right]+\frac{\partial}{\partial x} F\left[\phi(x,t)\right]=0 \tag{19.3}$$

for *all* solutions of (19.1), then (19.3) is said to be a *conservation law*, where $D[\cdot]$ is the conserved "density" and $F[\cdot]$ is the conserved "flow or flux." The functional

$$I\left[\phi\right]\equiv\int_{-\infty}^{\infty} D\left[\phi(x,t)\right]dx \tag{19.4}$$

is a constant of the motion provided the integral exists and the integrand satisfies the appropriate boundary condition at $x=\pm\infty$. A standard procedure is to determine a set of conservation laws and then use (19.4) to obtain constants of the motion.

One way to generate conservation laws is to employ an *infinitesimal Bäcklund* transformation.[14] We illustrate the construction of an infinite number of conservation laws[15] using the sine–Gordon equation

$$\phi_{xt}=\sin\phi, \tag{19.5}$$

together with its classical Bäcklund transformation

$$\psi_x=\phi_x+2a\sin\tfrac{1}{2}(\psi+\phi), \tag{19.6a}$$

$$\psi_t=-\phi_t+2a^{-1}\sin\tfrac{1}{2}(\psi-\phi). \tag{19.6b}$$

We assume a to be a small real number and seek $\psi(x,t,a)$ generated from ϕ through the infinitesimal transformation

$$\psi(x,t,a)\approx\sum_{i=0}^{\infty}\psi_i(x,t)a^i \quad \text{as } a\to 0. \tag{19.7}$$

Inserting (19.7) into (19.6b) yields

$$\sum_{i=0}^{\infty}\psi_{it}a^i=-\phi_t+\frac{2}{a}\sin\left[\frac{1}{2}\left(\sum_{i=0}^{\infty}\psi_i a^i-\phi\right)\right].$$

As $a\to 0$ this relation requires $\psi_0=\phi$ and $\psi_1=2\phi_t$. Upon calculating time deriva-

---

[14] C. Loewner, *Generation of solutions of systems of partial differential equations by composition of infinitesimal Bäcklund transformations*, J. Analyse Math., 2 (1953), p. 219.

[15] The KdV equation, nonlinear Schrödinger equation, and generalized sine–Gordon equation also have an infinite number of conservation laws.

tives, letting $a \to 0$, and equating the coefficients for higher powers of a, we have

$$\begin{aligned} &\psi_0=\phi, \quad \psi_1=2\phi_t, \quad \psi_2=2\phi_{tt}, \\ &\psi_3=2\phi_{ttt}+\tfrac{1}{3}\phi_t^3, \qquad \psi_4=2\phi_{tttt}+2\phi_t^2\phi_{tt} \\ &\psi_5=2\phi_{ttttt}+3\phi_t^2\phi_{ttt}+5\phi_t\phi_{tt}+\tfrac{3}{20}\phi_t^5 \\ &\cdots\cdots\cdots\cdots\cdots\cdots\cdots\cdots \end{aligned} \tag{19.8}$$

Some algebra establishes that this series for ψ is consistent with (19.6a).

From any fixed conservation law the aforementioned series enables us to derive an infinite number of conservation laws. The Lagrangian density for (19.5) is

$$\mathcal{L}=\tfrac{1}{2}\phi_x\phi_t-\cos\phi+1 \tag{19.9}$$

—i.e., (19.5) can be written in *conservation form* as

$$\left(\tfrac{1}{2}\psi_x^2\right)_t+(\cos\psi-1)_x=0. \tag{19.10}$$

Since (19.9) is symmetric in x and t, a corresponding expression with x and t interchanged can be used in place of (19.10). Substituting (19.7) into (19.10) and equating like even powers of a gives an infinite set of conservation laws whose first few densities are

$$\begin{aligned} D_0&=\tfrac{1}{2}\phi_x^2, \qquad D_2=2\phi_x\phi_{xtt}+2\phi_{xt}^2 \\ D_4&=2\phi_x\phi_{xtttt}+4\phi_{xtt}\phi_{xttt}+2\phi_{xtt}\phi_{xt}\phi_t^2 \\ &\quad+2\phi_t^2\phi_x\phi_{xtt}+2\phi_{xtt}^2+4\phi_x\phi_t\phi_{xt}\phi_{tt}. \end{aligned} \tag{19.11}$$

The equating of odd powers of a generates a set which is not independent, but merely contains the time derivatives of (19.11).

The first few densities for the KdV equation

$$\phi_t=6\phi\phi_x-\phi_{xxx}$$

are

$$\begin{aligned} &D_1=\phi, \quad D_2=\tfrac{1}{2}\phi^2, \quad D_3=-\phi_x^2+\tfrac{1}{3}\phi^3, \\ &D_4=\tfrac{1}{4}\phi^4-3\phi\phi_x^3+\tfrac{9}{5}\phi_{xx}^2, \\ &\cdots\cdots\cdots\cdots\cdots\cdots\cdots\cdots \end{aligned}$$

Conservation laws provide simple and efficient methods for the study of many qualitative properties of solutions, including stability, evolution of solitons, and decomposition into solitons, as well as the theoretical description of solution manifolds. There seems to be a close relationship between the existence of a sequence of conservation laws and the existence of solitons. Research on this relationship is in progress.

§20. Weakly Dispersive Shallow-Water Waves in Two Space Dimensions[16]

Some extensions of the Bäcklund transformation to problems in higher dimension are beginning to appear. Here we shall describe one such example concerned with weakly dispersive nonlinear shallow-water waves in two space dimensions (x, y). The procedures resemble those in one space dimension. However, the case in two dimensions contains many more solutions which are not directly related to solitons.

Consider the equation

$$q_{xt}+q_{yy}+q_{xx}+(3q^2)_{xx}+q_{xxxx}=0, \tag{20.1}$$

which has been used by Kadomtsev and Petviashvili[17] to describe disturbances in a weakly dispersive, weakly nonlinear medium. A Bäcklund transformation for (20.1) is $(w_x=q)$

$$\begin{aligned}&(w^1-w)^2+2(w^1+w)_x-\varepsilon\frac{2}{\sqrt{3}}\int^x(w^1-w)_y\,dx=c,\\ &4(w^1-w)_{xx}+(w^1-w)^3+3(w^1+w)_x(w^1-w)+3(w-w^1)_{xx}\\ &\qquad+\sqrt{3}\,\varepsilon(w+w^1)_y+(w^1-w)_x+\int^x(w^1-w)_t\,dx=0,\end{aligned} \tag{20.2}$$

where $\varepsilon=\pm 1$ is a result of the fact that we can have waves propagating in both the positive and negative y-directions.

From (20.2) we can proceed to construct specific solutions of (20.1) starting with a known solution, for example with $w=0$. The advantage of the Bäcklund transformation is the possibility of a nonlinear superposition for the solutions. Using a known solution w_0 with the first equation of (20.2) it is easily shown by our already demonstrated procedure that

$$w_3=w_1+w_2-w_0+\frac{2(w_1-w_2)_x}{w_1-w_2} \tag{20.3}$$

is the superposition sought.

[16]H. H. Chen, *A Bäcklund transformation in two dimensions*, submitted.

[17]See Chen, op. cit.

§21. Some Miscellaneous Applications

In this concluding section we list additional applications of the Bäcklund transformation and references of substance for the interested to peruse:

1. Some second-order nonlinear ordinary differential equations from physics.[18]
2. Two methods for integrating Monge-Ampère's equations.[19]
3. Study of invariant transformations for the hodograph equations[20] and other applications in gas dynamics.[21]
4. The role of groups in numerical analysis.[22,23]

---

[18]R. L. Anderson and J. W. Turner, *A type of Bäcklund-like invariance transformation for a class of second order ordinary differential equations*, Lett. Math. Phys., 1 (1975), pp. 37–42. In addition see J. W. Turner's contribution under the same title in the Proc. Internat. Joint IUTAM/IMU Symposium GROUP THEORETICAL METHODS IN MECHANICS, Novosibirsk, Inst. of Hydrodynamics, Siberian Branch of the USSR Academy of Science, Novosibirsk (1978), to appear, which incorporates the results contained in A. S. Fokas and R. L. Anderson, *Group nature of Bäcklund transformations*, Lett. Math. Phys., to appear.

[19]M. Matsuda, [1], *Two methods of integrating Monge-Ampère's equations*, Trans. Amer. Math. Soc., 150(1970), p. 327.

[20]G. Power, C. Rogers, and R. A. Osborn, *Bäcklund and generalized Legendre transformations in gas dynamics*, Z. Angew. Math. Mech., 6 (1969) p. 333.

[21]Loewner, op. cit.

[22]N. N. Janenko and Ju. I. Šokin, *Group classification of difference schemes for a system of one dimensional equations of gas dynamics*, Amer. Math. Soc. Transl. (2), 104 (1976), p. 259.

[23]W. F. Ames and N. H. Ibragimov, *Utilization of group properties in computation*, Proc. Internat. Joint IUTAM/IMU Symposium GROUP THEORETICAL METHODS IN MECHANICS, Novosibirsk, Inst. of Hydrodynamics, Siberian Branch of the USSR Academy of Science, Novosibirsk (1978), to appear.

References

W. F. AMES

[1] *Nonlinear Partial Differential Equations in Engineering*, vol. II, Academic Press, New York–San Francisco–London, 1972.

R. L. ANDERSON, A. O. BARUT, and N. H. IBRAGIMOV

[1] *Group Equivalence of Quantum Mechanical Systems*, unpublished.

R. L. ANDERSON, R. DI FRANCO, and D. SMITH

[1] $SL(n+2, \mathbb{R})$*-maximal space-time Lie invariance transformation group for a Newtonian free particle with n degrees of freedom*, unpublished.

R. L. ANDERSON and N. H. IBRAGIMOV

[1] *Bianchi-Lie, Bäcklund, Lie-Bäcklund transformation*, Proc. Internat. Joint IUTAM/IMU Symposium GROUP THEORETICAL METHODS IN MECHANICS, Novosibirsk, Inst. of Hydrodynamics, Siberian Branch of the USSR Academy of Science, Novosibirsk (1978), to appear.

R. L. ANDERSON, and M. MERNER

[1] *Transitive group structure for the one-body Keplerian problem*, Nonlinear Analysis, Theory, Methods and Applications, 2 (1978), pp. 627–634.

R. L. ANDERSON and D. PETERSON

[1] *Algorithm for global linearization and group theoretical equivalence of completely integrable $2n$-dimensional dynamical systems defined on* $\mathbb{R}^{2n}$, Nonlinear Analysis, Theory, Methods, and Applications, 1 (1977), pp. 481–493.

R. L. ANDERSON, T. SHIBUYA, and C. E. WULFMAN

[1] *On the connection between constants of the motion of different quantal systems*, Rev. Mexicana Fis., 23 (1974), pp. 257–272.

A. V. BÄCKLUND

[1] *Einiges über Curven- und Flächen-Transformationen*, Lunds Universitets Års-skrift, X, För Ar 1873, II. Afdelningen för Mathematik och Naturetenskap (1873–74), pp. 1–12.

[2] *Ueber Flächentransformationen*, Math. Ann. IX (1876), pp. 297–320.

[3] *Zur Theorie der partiellen Differentialgleichungen erster Ordnung*, Math. Ann. XVII (1880), pp. 285–328.

[4] *Zur Theorie der Flächentransformationen*, Math. Ann., XIX (1882), pp. 387–422.

[5] *Om ytor med konstant negativ krökning*, Lunds Universitets Årsskrift, 19 (1883).

A. O. BARUT

[1] *Conformal group→Schrödinger group→dynamical group—the maximal kinematical group of the massive Schrödinger particle*, Helv. Phys. Acta, 46 (1973), pp. 496–503.

A. O. BARUT and R. RĄCZKA

[1] *Theory of Group Representations and Applications*, PWN, Warsaw, 1977, Chapter 13, §4.

L. BIANCHI

[1] *Ricerche sulle superficie a curvatura costante e sulle elicoidi*, Ann. Scuola Norm. Sup. Pisa, II (1879), p. 285.

[2] *Lezioni di Geometria Differenziale*, vol. I, Enrico Spoerri, Pisa, 1922, pp. 743–747.

G. BIRKHOFF and S. MACLANE

[1] *A Survey of Modern Algebra*, (revised) MacMillan, New York, 1953, pp. 293–298.

J. CLAIRIN

[1] *Sur les Transformations de Baecklund*, Ann. Sci. Ecole Norm. Sup. (3), (1902), supplément, pp. 1–63.

G. DARBOUX

[1] *Leçons sur la theorie générale des surfaces*, Vol. III, Gauthier-Villars et Fils, Paris, 1894, pp. 438–444.

P. A. M. DIRAC

[1] *The Principles of Quantum Mechanics*, Clarendon Press, Oxford, 4th ed., 1958, p. 248.

L. P. EISENHART

[1] *A Treatise on the Differential Geometry of Curves and Surfaces*, Dover Publications, New York, 1960.

V. FOCK

[1] *Zur Theorie des Wasserstoffatoms*, Z. Phys. 98 (1935), pp. 145–154.

W. H. GOODYEAR

[1] *Completely general closed-form solution for coordinates and partial derivatives of the two-body problem*, Astronomical J. 70 (1965), p. 189.

E. GOURSAT

[1] *Leçons sur l'intégration des equations aux dérivées partielles du second ordre*, 1902, *vol. II, p.* 248.

[2] *Le Problème de Bäcklund*, Mémorial des sciences mathématiques, Fasc. VI, Gauthier-Villars, Paris, 1925.

J. HADAMARD

[1] *Lectures On Cauchy's Problem in Linear Partial Differential Equations*, Yale University Press, New Haven, 1923.

S. HERRICK

[1] Astrodynamical Rept. No. 7, ASTIA Document AD 250 757, ¶6Q (1960).

N. H. IBRAGIMOV

[1] *Group properties of some equations of physics*, Institute of Hydrodynamics, USSR Academy of Sciences, Siberian Branch, Novosibirsk, 1967. (In Russian.)

[2] *Group properties of some differential equations*, "Nauka" Siberian Branch, Novosibirsk, 1967. (In Russian.)

[3] *Invariant variational problems and conservation laws*, Theoret. and Math. Phys., 1 (1969), pp. 350–359. (In Russian.)

[4] *Lie Groups In Some Problems of Mathematical Physics*, Novosibirsk Univ. 1972. (In Russian.)

[5] *Invariance and conservation laws of continuum mechanics*, Proc. Symposium on Symmetry, Similarity and Group Theoretic Methods in Mechanics, Calgary, P. G. Glockner and M. C. Singh, eds., Univ. of Calgary, Calgary (1974), pp. 63–82.

[6] *Lie-Bäcklund Groups and Conservation Laws*, Dokl. Akad. Nauk SSSR, 230 (1976), no. 1, Soviet Math. Dokl., 17 (1976), pp. 1242–1246.

[7] *Group theoretical nature of conservation laws*, *Lett. Math. Phys.*, 1 (1977), pp. 423–428.

[8] Unpublished.

N. H. IBRAGIMOV and R. L. ANDERSON

[1] *Lie-Bäcklund tangent transformations*, Math. Anal. Appl., 59 (1977), pp. 145–162.

F. Klein

[1] *Über die Differentialgesetze für die Erhaltung von Impuls und Energie in der Einsteinschen Gravitationstheorie*, Nachr. Ges. Wiss. Göttingen Math.-Phys. Kl. (1918), Heft 2, pp. 171–189.

S. Lie

[1] *Begründung einer Invarianthentheorie der Berührungstransformationen*, Math. Ann. VIII (1874), Heft 3, pp. 215–288 (see Note, p. 223).

[2] *Zur analytischen Theorie der Berührungstransformationen*, Kristiania Forh. Aaret (1874), pp. 237–262.

[3] *Zur Theorie der Flächen konstanter Krümmung* III, Arch. Math. og Naturvidenskab, V (1880), Heft 3, pp. 282–306.

[4] *Zur Theorie der Flächen konstanter Krümmung* IV, Arch. Math. og Naturvidenskab, V (1880), Heft 3, pp. 328–358.

[5] *Gesammelte Abhandlungen*, B. G. Teubner, Leipzig: vol. 3, 1922; vol. 4, 1924; vol. 6, 1927.

[6] *Vorlesungen über continuerliche Gruppe*, Chelsea Publishing Co., New York, 1971. (First published Leipzig, 1893.)

[7] *Über die Integration durch bestimme Integrale von einer Klasse linearer partieller Differentialgleichungen*, Arch. for Math., 6 (1881), Heft 3, pp. 328–368.

V. Niederer

[1] *The maximal kinematical invariance group of the free Schrödinger equation*, Helv. Phys. Acta 45 (1972), pp. 802–810.

E. Noether

[1] *Invariante Variationsprobleme*, Nachr. Ges. Wiss. Göttingen Math.-Phys. Kl. (1918), Heft 2, pp. 235–257.

L. V. Ovsjannikov

[1] *Groups and group invariant solutions of differential equations*, Dokl. Akad. Nauk SSSR, 118 (1958), no 3, pp. 439–442.

[2] *Group Properties of Differential Equations*, Siberian Branch of the USSR Academy of Sciences, Novosibirsk, 1962. (In Russian.)

[3] *Analytical Groups*, Novosibirsk Univ., 1972 (In Russian.)

[4] *Group Analysis of Differential Equations*, Nauka, Moscow, 1978. (In Russian.)

K. Stumpff

[1] *Calculation of ephemerides from initial values*, NASA Technical Note D-1415 (1962).

Index